Finite Element Analysis
Theory and Practice

Finite Element Analysis
Theory and Practice

M J FAGAN
University of Hull

Longman
Scientific &
Technical

Longman Scientific and Technical,
Longman Group UK Limited,
Longman House, Burnt Mill, Harlow,
Essex CM20 2JE, England
and Associated Companies throughout the world.

*Copublished in the United States with
John Wiley & Sons, Inc., 605 Third Avenue, New York, NY 10158*

© Longman Group UK Limited 1992

First published 1992

British Library Cataloguing in Publication Data
Fagan, M.J.
   Finite element analysis.
   I. Title
   620.001
ISBN 0–582–02247–9

Library of Congress Cataloging-in-Publication Data

Fagan, M.J., 1957–
   Finite element analysis / M.J. Fagan.
     p.   cm.
   Includes bibliographical references and index.
   ISBN 0–470–21817–7
   1. Finite element method.   I. Title
   TA347.F5F34 1992
   620′.001′51535—dc20

91–42924
CIP

Set by 8 in 10/13 Times
Produced by Longman Singapore Publishers (Pte) Ltd.
Printed in Singapore

For my wife, Gay, and daughter, Kate

List of Chapters

Contents

Preface

This book is an elementary text on the finite element method. It is aimed at engineering and science undergraduates with no previous knowledge of the method, and deliberately attempts to keep the mathematics of the subject as straightforward as possible. It is assumed that the reader does understand the basic concepts and equations of elasticity and thermal heat flow, and is familiar with simple matrix algebra.

The finite element method has developed into a sophisticated and apparently complex technique, but the fundamental principles are simple and easily understood if treated in a direct and logical manner. The procedure may be introduced in two different ways. Firstly, it may be presented as a mathematical tool, with the theory developed as a general procedure for obtaining approximate solutions to elliptic partial differential equations. Alternatively, the method may be introduced less rigorously, aimed specifically at those areas where it is most commonly used. If a novel area is encountered, or the theory needs to be developed further, the reader may then study more advanced texts, equipped with a sound knowledge of the fundamental principles. Since this second approach also allows the practical aspects of the method to be introduced more effectively, it is adopted here.

It cannot be overemphasized that the finite element method is an approximate method of analysis, and that the way an object is modelled, such as the choice of element or the representation of the loading and constraint conditions, is vital. Furthermore, considerable engineering judgement may be required to analyse the very detailed information that a finite element model can produce. Consequently, although sophisticated commercial finite element packages are available, the theory of finite elements and the practical (modelling and analysis) aspects of the technique must be clearly understood by the user, if the model he or she develops is to be valid and accurate.

In many cases an engineer will concentrate on either field problems (usually thermal) or elasticity problems, and for this reason the two areas are not generally intermixed in this text. Each is dealt with in separate chapters, and if not required the unwanted chapters may be omitted. For this reason, the chapters before and after the specialized ones tend to be general, with advanced applications of the finite element method covered later in the book.

Thus this text attempts to introduce the finite element method in a simple manner, and places particular emphasis on the practical side of the procedure. It does not discuss how the method is programmed into the computer, which usually need not concern the engineer. However, those features of the program which affect the efficient running and execution of the analysis are highlighted.

A brief summary of the contents of each chapter is now included to allow the reader to obtain an overview of the whole book:

Chapter 1 discusses the capabilities of the finite element method, and lists the types and classes of problems that the method can be used to analyse.

Chapter 2 introduces the basic principles behind the finite element method, and demonstrates the technique with a complete and yet simple one-dimensional model and an analysis of both a stress and a thermal problem.

Chapter 3 is the first chapter on practical aspects, and deals with the discretization of the object being modelled, the division into elements, and the location and optimal numbering of the nodes. Modelling economies through the use of model symmetry are also introduced.

Chapter 4 discusses the basic concepts and properties behind some linear elements, *i.e.* the choice of approximating function and the use of natural coordinate systems.

Chapter 5 covers stress analysis problems, and describes the formulation of the general element matrices and vectors for elasticity problems (by minimization of potential energy). One-, two- and three-dimensional and axisymmetric problems are examined.

Chapter 6 covers field problems, and describes the formulation of the general element matrices and vectors for field problems (by minimization of a functional and the weighted residual method). Thermal problems and the torsion of shafts are discussed in detail.

Chapter 7 demonstrates how the element equations are assembled to give the system equations, the boundary conditions are applied, and the equations are then solved for the unknown nodal variables.

Chapter 8 introduces more sophisticated elements with higher-order approximating functions, and the more complex techniques that they demand, *i.e.* the transformation between natural and global coordinate systems and the use of numerical integration. It also describes structural beam and plate elements.

Chapter 9 is another chapter on practical aspects. It **deals with model validity and accuracy** (*i.e.* how reliably the physical problem is modelled and how close the model is to convergence), together with the effect of element selection and mesh refinement, and element distortion. Two final sections discuss how the finite element results can be processed and analysed, and how they can and should be checked.

Chapter 10 **introduces more advanced applications of the finite element method.** In particular, it illustrates how problems with geometrical and material non-linearities can be investigated, together with the analyses of buckling and dynamic problems.

Chapter 11 deals further with the practical side of the method. **It discusses problem symmetry in more detail** and **introduces the ideas of submodelling and substructuring**, and describes element and program validation by patch and benchmark tests.

Chapter 12 **discusses commercially available finite element packages**, the type of facilities that can be expected from them, and in particular the sophisticated pre- and post-processors on which most of them rely.

Although a simple finite element problem can be introduced and worked through in just a few pages (in Chapter 2), the finite element method is a 'long' procedure; there are many steps in a complete analysis. The new student of the method can become bewildered as to where each step fits in the overall scheme, and for this reason a figure is included here to show the general flow of the technique. The layout of the first half of this book follows the outline of the method in the figure, with more detail and more advanced topics in the second half.

Many worked examples are included through the book to introduce the different aspects of the theory. These examples tend to be simple, so that the principles are presented without filling the pages with complex and usually repetitive calculations. Whether a model uses four elements or forty, the analysis proceeds in the same way. The difference is in the final system equations, with the latter producing many times more equations that cannot reasonably be processed by hand, but can easily be processed by computer. It should be noted, however, that these examples are an integral part of the text and should be read with the rest of the chapter in which they appear. Each contains comments not only about the specific problem, but also about the general procedures and implications of the results of the example.

Finally, a substantial glossary is incorporated at the end of the book to help the student become familiar with and understand the many specialized terms used in this subject.

1. Discretization of the problem.
The geometry is divided into a finite number of regions or elements, (taking account of any symmetry, different matrials, loading and boundary conditions.

2. Selection of the approximating function.
The form of the approximating function (for the displacement or temperature) is selected.

Simplex element.
Linear function.

Higher order elements.
Quadratic or cubic functions.

3. Derivation of the basic element equations.
The equations to describe the behaviour of the selected element type are derived.

Variational formulation method.

Weighted residual method.

4. Calculation of the system equations.
a. The individual element equations are calculated.

Simplex elements.
Exact evaluation.

Higher order elements.
Evaluation using numerical integration.

b. The element equations are combined to give the system equations.

5. Encorporation of the boundary conditions.
The system equations are modified to take account of the constraints.

6. Solution of the system equations.
The system equations are solved to give the nodal displacements or temperartures.

Elimination method.

*Wavefront method. ***

7. Post-processing.
The strains, stresses or heat flows etc. are calculated as required.

Simplex element.
Evaluation at nodes.

Higher order elements
Evaluation at integration points.
Extrapolation to nodes.

8. Presentation of results.
a. The nodal values are averaged.
b. The results are printed or plotted.

pre-processing phase

analysis phase

post processing phase

Main steps of the finite element method

[* in this method steps 4,5 and 6 are not discrete.]

Notation

Matrix and vector notation

The presentation of the finite element method requires the use of matrix and vector notation, and in this text such matrices and vectors are identified by brackets, rather than bold-face type used in many textbooks, such as **U**, **N**, **B**. This has the advantage that the equations can be more easily written down and manipulated by hand, but it also facilitates their presentation in computer-based learning methods and programs. In particular, the following schemes are used throughout the book.

1. Column vectors are enclosed in curly brackets, for example,

$$\{U\} = \begin{Bmatrix} u_1 \\ u_2 \\ u_3 \\ u_4 \end{Bmatrix}$$

2. Row vectors and matrices are enclosed in square brackets, for example,

$$[N] = [N_1 \; N_2 \; N_3 \; N_4]$$

$$[B] \quad \begin{bmatrix} b_i & b_j & b_k & b_l \\ c_i & c_j & c_k & c_l \\ d_i & d_j & d_k & d_l \end{bmatrix}$$

and also, therefore, $\{U\}^{\mathrm{T}} = [u_1 \; u_2 \; u_3 \; u_4]$

Mathematical symbols

The following is a list of the mathematical symbols used in this text. A few symbols have more than one meaning in different parts of the book, but this should not cause any ambiguity. Other symbols which occur in particular limited pieces of analysis or examples are defined as they are used. Many of them appear with subscripts or superscripts, and possibly both, and these are detailed at the end of the list.

A area

$[B]$ matrix relating strain vector to displacement, or thermal gradient vector to temperature

C damping constant (in dynamic problems)

$[C]$ structural damping matrix (in dynamic problems)

$[C]$ specific heat matrix (in transient thermal problems)

$[D]$ property matrix

e general element

E Young's modulus

$\{F\}$ force vector

$\{g\}$ thermal gradient vector

G shear modulus

h convection coefficient

H length of element side

H_i weighting function for numerical integration

I area moment of inertia

$[J]$ Jacobian matrix

$[k]$ stiffness matrix

$[k_G]$ geometric stiffness matrix (in non-linear problems)

K thermal conductivity

l direction cosine with respect to the x axis

L length

L_n natural coordinate ($n = 1, 2, ...$)

m direction cosine with respect to the y axis

M torque, bending moment

$[M]$ mass matrix

n direction cosine with respect to the z axis

N_n shape function ($n = i, j, k$ or $1, 2, 3, ...$)

$[N]$ shape function matrix

p pressure

P perimeter length (in field problems)

P applied nodal force (in stress problems)

$\{P\}$ nodal force vector (in stress problems)

q heat flow rate or flux

Q applied heat source or sink

R reaction (at a constraint)

$\mathbb{R}$ body force per unit volume in the r direction

S surface

S shear force (in beam, plate and shell elements)

t	thickness
t	time (in dynamic and transient problems)
T	temperature
u	displacement in the x direction
$\{U\}$	displacement vector
v	displacement in the y direction
v_{g}	ground acceleration (in dynamic problems)
V	volume
w	displacement in the z direction
W	work done
W_B	work done by body forces
W_P	work done by pressure loads
W_C	work done by nodal loads
$\mathbb{X}$	body force per unit volume in the x direction
$\mathbb{Y}$	body force per unit volume in the y direction
$\mathbb{Z}$	body force per unit volume in the z direction

Greek symbols

α	coefficient of thermal expansion
α_n	constant used in interpolation functions ($n = 1, 2, 3, \ldots$)
γ	shear strain
ΔT	change in temperature
ε	direct strain
$\{\varepsilon_0\}$	initial strain vector
$\{\varepsilon\}$	strain vector
θ	angle of rotation
θ_1	twist per unit length (in torsion problems)
λ	eigenvalue (in buckling problems)
$[\lambda]$	coordinate transformation matrix
Λ	strain energy
ν	Poisson's ratio
ξ	modal damping ratio (in dynamic problems)
Π	potential energy
σ	direct stress
$\{\sigma\}$	stress vector
τ	shear stress
$\{v\}$	mode shape (in dynamic problems)
Υ	general coordinate (in dynamic problems)
ϕ	unknown function, temperature, Prandtl's stress function
ϕ_∞	surrounding ambient temperature
$\{\Phi\}$	unknown function vector
ω	natural frequency

General

Subscripts

i,j,k	refers to general nodes i, j or k etc.
s	refers to the whole system or model
r,θ,z	refers to directions r, θ or z
x,y,z	refers to directions x, y or z
$1,2,3$	refers to nodes 1, 2 or 3 etc.

Superscripts

(e)	refers to general element (e)
o	refers to the global coordinate system
T	transpose of a vector or matrix

Special

$\bar{X}$	overscored – average value of X
$\underline{X}$	underlined – X at the centroid of an element
$\dot{X}$	derivative – first derivative of X with respect to time
$\ddot{X}$	derivative – second derivative of X with respect to time

Coordinate systems

$\xi\eta$	natural coordinate system
$r\theta z$	cylindrical coordinate system
xyz	Cartesian coordinate system

1   Background and application

The finite element method is not a new technique; it was first introduced in the 1950s, and has been continually developed and improved since then. It is now an extremely sophisticated tool for solving numerous engineering problems and is widely used and accepted in many branches of industry. Its development has not been paralleled by any other numerical analysis procedure, and it has made many other numerical analysis techniques and experimental testing methods redundant. For example, in the car industry, the structural integrity and performance of any new car design is thoroughly analysed and evaluated with finite element models, possibly years before the first prototype is built. The method is used to analyse the strength of individual components and that of the car as a whole, its impact behaviour and crashworthiness, the frequency characteristics of the total structure and its separate components, the temperature distribution in the engine block and pistons and the resulting thermal stresses, and many other areas.

Aircraft and aerospace companies, like many other industries, are similarly dependent on finite element analyses. The efficient design of any modern aeroplane is impossible without the technique. Indeed, many aircraft components and hence the total machine are certified and given airworthiness certificates through the results of finite element models.

Clearly the areas of application and the potential of the finite element method are enormous. The growth of the technique is attributable directly to the rapid advances in computer technology and computing power, particularly over the last decade. As the power of computers has increased, so it has been possible to analyse larger and more complex problems. Combined with this effect, the decrease in price and increase in availability of moderately powerful mini- and microcomputers has meant that even small companies can have access to finite element programs.

The number and size of software companies developing and supporting commercial finite element packages has grown to meet the demand for the programs, and it is now a multimillion pound industry itself.

There are three broad problem areas that can be investigated by the finite element method. These are introduced below with examples illustrating the wide range of applications of the technique.

Steady state problems

Steady state or equilibrium analyses are the most common use of the finite element method. For elasticity problems, a body under equilibrium conditions can be analysed and its distortion predicted. From the calculated values of displacement, it is then possible to derive the strains and stresses experienced by the body. Thermal analyses are also frequently performed by the method; the temperature distribution and heat flow through a body can be predicted for a wide variety of boundary conditions. Table 1.1 lists **some** of the areas where the method is applied to equilibrium problems.

Eigenvalue problems

Eigenvalue problems are an extension of equilibrium problems, but involve the calculation of fundamental characteristics of the body or system under examination. For example, the method can be used to determine the natural frequencies and mode shapes of components, and the buckling loads of structures. Samples of the type of eigenvalue

Table 1.1. Typical equilibrium problems suitable for analysis by the finite element method

Area	Typical application
Aerospace engineering	Stress analysis of aircraft frames, wings, missile and spacecraft components; thermal analysis of gas turbine blades, heat exchangers
Automotive engineering	Stress analysis of crankshaft, cylinder block, connecting rods, chassis; thermal analysis of pistons; lubrication of big-end bearings
Biomedical engineering	Stress analysis of bones, hip replacements, teeth and heart
Civil engineering	Stress analysis of dams, retaining walls, excavations; soil mechanics
Electrical engineering	Steady state thermal analysis of integrated circuit boards
Hydraulic engineering	Analysis of water seepage and flow under dams; aquifer analysis
Mechanical engineering	General one-, two- and three-dimensional and axisymmetric stress analyses of components; stress analysis of shafts, gears and pressure vessels; crack propagation
Nuclear engineering	Stress analysis of reactor vessels and structures; thermal analysis of reactor components
Structural engineering	Static analysis of electricity pylons, girders and bridges

problems that can be analysed by the finite element method are presented in Table 1.2.

Transient problems

In eigenvalue problems, time does not appear explicitly, although the natural frequency characteristics of a body might be determined. In transient or propagation problems, however, the loads can be functions of time, and the finite element method is used to calculate the forced response of the body. The propagation of stress waves and transient heat flows is also considered under this heading. Other examples are included in Table 1.3.

The range of problems suitable for analysis by the finite element method is clearly large, and certainly many of these problems were previously

Table 1.2. Typical eigenvalue problems suitable for analysis by the finite element method

Area	Typical application
Aerospace engineering	Frequency analysis of engine components, helicopter rotor blades, gearbox casing; acoustic analysis of aircraft passenger compartments
Automotive engineering	Acoustic analysis of passenger compartment and exhaust system; frequency analysis of gearbox casing and body shell
Electrical engineering	Natural frequencies of printed circuit boards
Hydraulic engineering	Natural periods of lakes and harbours; sloshing of liquids
Mechanical engineering	Natural frequency of components, shafts; critical buckling loads
Nuclear engineering	Neutron flux distribution; frequency analysis of pressure vessels
Structural engineering	Natural frequency and buckling loads of structures; vibration analysis of multistorey buildings

Table 1.3. Typical propagation problems suitable for analysis by the finite element method

Area	Typical application
Aerospace engineering	Forced frequency analysis of aircraft and spacecraft components; thermal analysis of rocket nozzles
Automotive engineering	Time dependent analysis of engine components, piston, disc brakes, exhaust; crashworthiness of chassis
Biomedical engineering	Impact analysis of skull; dynamic analysis of body and limbs
Civil engineering	Stress waves in rock structures
Hydraulic engineering	Transient seepage and flow
Mechanical engineering	Analysis of impact problems; dynamic crack propagation
Nuclear engineering	Time dependent thermal analysis of reactor components; shock spectrum analyses
Structural engineering	Shock and earthquake analysis of buildings and bridges

insoluble until the finite element method became available. Increasingly, the complexities and approximations of the algorithms employed by the many commercial finite element packages that are now widely used are becoming less apparent to the user. The essential procedures of the method are invariably concealed by sophisticated pre- and post-processors. These interfaces with the program have become more important as the cost of the software and hardware has decreased while labour costs have increased. The limiting factor on the use of the finite element method in most companies is now the manpower, rather than the computing power as was the case just a few years ago.

The development of the pre- and post-processors has not only made the user remote from the method, but also given the programs an appearance of unquestionable accuracy. Pre-processors can now easily generate complex, visually impressive and apparently reliable models with a minimum amount of input from the engineer, while the post-processors produce equally impressive and convincing graphical output. The finite element method can produce accurate and reliable results when used correctly, but it is important to remember that it is an approximate technique, and the validity and accuracy of the model and its solution rely on an accurate representation of the problem and correct analysis procedures.

As a result of the enormous growth in the number of finite element users and commercial software packages, a body called the National Agency for Finite Element Methods and Standards (NAFEMS) was established in the UK in 1983. The primary functions of the Agency are to assist in the development of error-free finite element programs and to give users guidance and general advice on the application of the programs. NAFEMS now publishes guidelines on finite element modelling, element validation tests and standards, and benchmark assemblies. The element validation work investigates how individual elements behave, particularly when they are distorted, while the benchmark tests compare the performance of element assemblies with known theoretical solutions. The work of NAFEMS and similar groups will become of increasing importance as finite element modelling in all branches of engineering continues to expand.

2 Introduction to the method

2.1 General theory

The basic principles underlying the finite element method are simple. Consider, for example, a body in which the distribution of an unknown variable (such as temperature or displacement) is required. Firstly, the region is divided into an assembly of subdivisions called elements, which are considered to be interconnected at joints, known as nodes (Fig. 2.1). The variable is assumed to act over each element in a predefined manner, with the number and type of elements chosen so that the variable distribution through the whole body is adequately approximated by the combined elemental representations. The distribution across each element may be defined by a polynomial (for example, linear or quadratic) or a trigonometric function.

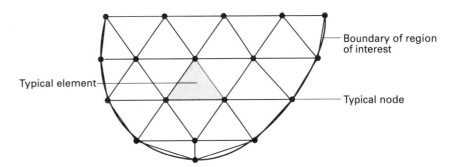

Figure 2.1 Discretization of a region into a number of finite elements

After the problem has been discretized, the governing equations for each element are calculated and then assembled to give the system equations. The element equations may be found in a variety of ways, as will be discovered later, but it turns out that the equations of a particular type of element for a specific problem area (stress or thermal, for example) have a constant format. Thus, once the general format of the equations of an element type is derived, the calculation of the equations for each occurrence of that element in the body is straightforward; it is simply a question of substituting the nodal coordinates, material properties and loading conditions of the element into the general format.

The individual element equations are assembled to obtain the system equations, which describe the behaviour of the body as a whole. These generally take the form

$$[k]\{U\} = \{F\} \qquad [2.1]$$

where $[k]$ is a square matrix, known as the stiffness matrix; $\{U\}$ is the vector of (unknown) nodal displacements or temperatures; and $\{F\}$ is the vector of applied nodal forces.

Equation 2.1 is directly comparable to the equilibrium or load–displacement relationship for a simple one-dimensional spring, where a force F produces a deflection U in a spring of stiffness k. To find the displacement developed by a given force, the relationship is inverted. The same approach applies to the finite element method; however, before Eq. 2.1 can be inverted and solved for $\{U\}$, some form of boundary condition must be applied. In stress problems, this means that the body must be constrained to prevent it from performing unlimited rigid body motion. For thermal problems, the temperature must be defined at one or more of the nodes.

The solution of Eq. 2.1 is not trivial in practice because the number of equations involved tends to be very large. It is not unreasonable to have 50 000 equations, and consequently $[k]$ cannot be simply inverted. Fortunately, however, $[k]$ is banded, and techniques have been developed to store and solve the equations efficiently. After solving for the unknown nodal values, it is then simple to use the temperatures to calculate the elemental heat flows, or the displacements to find the strains and then the elemental stresses.

Thus the finite element method is a straightforward and logical procedure following a well-defined path. Two simple one-dimensional examples are now introduced to show how the method can be used to predict the stresses in a stepped bar, and the thermal distribution through a wall.

2.2   A simple one-dimensional element: the pin-jointed bar

The simplest element to introduce the finite element method in stress analysis is the one-dimensional bar, as shown in Fig. 2.2(a). Each end of the bar is assumed to be pin-jointed, so that it can only transmit a tensile or a compressive axial force. The basic finite element corresponding to such a bar is presented in Fig. 2.2(b). It has two nodes, i and j; it lies along the x axis, and consequently only experiences an axial displacement. Assuming that the displacement u varies linearly along the length of the bar L, then

$$u = a + bx \qquad [2.2]$$

where a and b are constants.

(a)

(b)

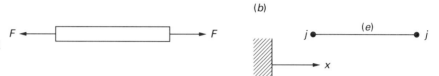

Figure 2.2 (a) One-dimensional pin-jointed bar (b) equivalent finite element

If u_i and u_j are the unknown displacements at each node, then

$$\begin{aligned} u_i &= a + bx_i \\ u_j &= a + bx_j \end{aligned} \qquad [2.3]$$

Since the coordinates of the nodes are known (x_i and x_j), these two equations can be solved for a and b, giving

$$\begin{aligned} a &= (u_i x_j - u_j x_i)/L \\ b &= (u_j - u_i)/L \end{aligned} \qquad [2.4]$$

Substituting these back into Eq. 2.2 and rearranging, it is found that

$$u = \frac{x_j - x}{L} u_i + \frac{x - x_i}{L} u_j \qquad [2.5]$$

or

$$u = N_i u_i + N_j u_j \qquad [2.6]$$

N_i and N_j are known as the shape functions of the element, and are a very important feature of any finite element. The relevance of shape functions is discussed in Chapter 4.

When a structure is loaded and attains an equilibrium position, its potential energy must be a minimum. The potential energy Π of the structure can be defined as

$$\Pi = \Lambda - W \qquad [2.7]$$

where Λ is the strain energy and W is the work done by any external loads.

For the single element considered here, the strain energy stored in the element following any deformation (assuming constant cross-sectional area A) is

$$\Lambda = \int_{x_i}^{x_j} \frac{1}{2} \sigma \varepsilon A \ dx \qquad [2.8]$$

However, the axial strain ε is related to the stress σ by Young's modulus E, so that

$$\Lambda = \frac{AE}{2} \int_{x_i}^{x_j} \varepsilon^2 \ dx \qquad [2.9]$$

The strain is simply defined as du/dx, which from Eq. 2.5 is seen to be

$$\varepsilon = (-u_i + u_j)/L \qquad [2.10]$$

Note that since a linear variation of u was chosen, the strain is constant over the element.

Substituting into Eq. 2.9 and integrating gives

$$\Lambda = \frac{AE}{2L} (-u_i + u_j)^2 \qquad [2.11]$$

It is more convenient to write this in matrix form as

$$\Lambda = \frac{AE}{2L} [u_i \ u_j] \begin{bmatrix} 1 & -1 \\ -1 & 1 \end{bmatrix} \begin{Bmatrix} u_i \\ u_j \end{Bmatrix} = \frac{1}{2} \{U\}^{\mathrm{T}}[k]\{U\} \qquad [2.12]$$

where

$$\{U\} = \begin{Bmatrix} u_i \\ u_j \end{Bmatrix} \quad \text{and} \quad [k] = \frac{AE}{L} \begin{bmatrix} 1 & -1 \\ -1 & 1 \end{bmatrix} \qquad [2.13]$$

and $[k]$ is called the stiffness matrix for the element.

For a bar element, the only external forces that can be applied are nodal forces (F_i and F_j) acting at the ends of the bar, so that the work done by external forces is

$$W = u_i F_i + u_j F_j = \{U\}^{\mathrm{T}}\{F\} \qquad [2.14]$$

Therefore, for the single bar, the total potential energy is

$$\Pi = \frac{1}{2}\{U\}^{\mathrm{T}}[k]\{U\} - \{U\}^{\mathrm{T}}\{F\} \qquad [2.15]$$

For minimum potential energy, the displacements must be such that

$$\frac{\partial \Pi}{\partial u_i} = \frac{\partial \Pi}{\partial u_j} = 0 \quad \text{or} \quad \frac{\partial \Pi}{\partial \{U\}} = 0$$

The differentiation of the terms in Eq. 2.15 is straightforward and is discussed in most textbooks covering matrix algebra. In this particular case, the answer is easily verified by performing the calculation with the expanded form. The result is

$$\frac{\partial \Pi}{\partial \{U\}} = [k]\{U\} - \{F\} = 0 \qquad [2.16]$$

and therefore

$$\left\{ \begin{array}{c} F_i \\ F_j \end{array} \right\} = \frac{AE}{L} \left[\begin{array}{cc} 1 & -1 \\ -1 & 1 \end{array} \right] \left\{ \begin{array}{c} u_i \\ u_j \end{array} \right\} \qquad [2.17]$$

Not surprisingly, this is the most general form of the load–displacement relationship for a spring, where a force is applied to either end of the spring. For example,

$$F_i = \frac{AE}{L}(u_i - u_j) = \frac{AE}{L} \times \text{extension}$$

where AE/L is the stiffness of the spring.

The derivation detailed above only considers one element. In practice a model will consist of many elements, and it should be remembered that it is the total potential energy of the system that must be minimized. Therefore if, for example, E bar elements are used, the total potential energy is

$$\Pi = \sum_{e=1}^{E} (\Lambda^{(e)} - W) \qquad [2.18]$$

where $\Lambda^{(e)}$ is the strain energy for element (e), and the work done W by the external loads is

$$W = u_1 F_1 + u_2 F_2 + \dots + u_n F_n = \{U\}^{\mathrm{T}} \{F\} \qquad [2.19]$$

if the model uses n nodes. So minimization gives

$$\frac{\partial \Pi}{\partial \{U\}} = \left(\sum_{e=1}^{E} [k^{(e)}] \right) \{U\} - \{F\} = 0 \qquad [2.20]$$

where the term $\sum_{e=1}^{E} ([k^{(e)}])$ is the global or system stiffness matrix, and is the sum of all the element system matrices.

A simple stress analysis of a stepped bar now follows, to demonstrate the use of these equations.

2.2.1 Stress analysis of a stepped bar

The axially loaded bar shown in Fig. 2.3(a) can be analysed by the basic element developed in the previous section. The finite element idealization is drawn in Fig. 2.3(b). Only two elements are chosen for this example, but naturally more could be used if thought necessary.

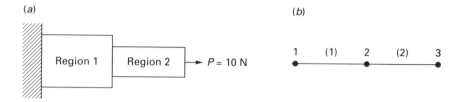

Figure 2.3 (a) Simple stepped bar for analysis (b) finite element model

$A^{(1)} = 20 \text{ mm}^2, A^{(2)} = 10 \text{ mm}^2$
$L^{(1)} = L^{(2)} = 100 \text{ mm}$
$E^{(1)} = E^{(2)} = 200 \times 10^3 \text{ MPa}$

The general element in the previous section is identified by nodes i and j. However, in the stepped bar the nodes are labelled numerically and consecutively, so that for each element they correspond as follows:

Element	i node	j node
(1)	1	2
(2)	2	3

This information is called the element connectivity. It is important because it contains details of how the elements are connected together, and how their stiffness matrices should be assembled into the global stiffness matrix.

So for element (1), Eq. 2.13 gives the stiffness matrix as

$$[k^{(1)}] = \frac{A^{(1)}E^{(1)}}{L^{(1)}} \begin{bmatrix} 1 & -1 \\ -1 & 1 \end{bmatrix} = \begin{bmatrix} 4 & -4 \\ -4 & 4 \end{bmatrix} \times 10^4 \text{ N/mm}$$

and for element (2),

$$[k^{(2)}] = \frac{A^{(2)}E^{(2)}}{L^{(2)}} \begin{bmatrix} 1 & -1 \\ -1 & 1 \end{bmatrix} = \begin{bmatrix} 2 & -2 \\ -2 & 2 \end{bmatrix} \times 10^4 \text{ N/mm}$$

Now the assembly of the system stiffness matrix must be performed, where each element matrix is added into the system matrix in turn. Element (1) has nodes 1 and 2, and its rows and columns are associated with u_1 and u_2 as follows:

$$[k^{(1)}] = \begin{matrix} & u_1 & u_2 & \\ & \begin{bmatrix} 4 & -4 \\ -4 & 4 \end{bmatrix} & \begin{matrix} u_1 \\ u_2 \end{matrix} \end{matrix} \times 10^4 \text{ N/mm}$$

Similarly, element (2) is associated with nodes 2 and 3, and its stiffness matrix can be labelled as

$$[k^{(2)}] = \begin{matrix} & u_2 & u_3 \\ \begin{bmatrix} 2 & -2 \\ -2 & 2 \end{bmatrix} & \begin{matrix} u_2 \\ u_3 \end{matrix} \end{matrix} \times 10^4 \text{ N/mm}$$

Thus the system matrix, which has a row and a column associated with each degree of freedom and is therefore of size 3×3, is filled by placing the terms from each element stiffness matrix into the correct location. Therefore

$$[k] = \begin{matrix} & u_1 & u_2 & u_3 \\ \begin{bmatrix} 4 & -4 & 0 \\ -4 & 4+2 & -2 \\ 0 & -2 & 2 \end{bmatrix} & \begin{matrix} u_1 \\ u_2 \\ u_3 \end{matrix} \end{matrix} \times 10^4 \text{ N/mm} \qquad [2.21]$$

No terms appear in the (u_1, u_3) locations, because no element connects these two degrees of freedom together.

The global force vector in Eq. 2.20 is the vector of applied nodal forces, that is

$$\{F\}^T = [0 \ 0 \ 10] \qquad [2.22]$$

since a force is only applied at node 3 (of 10 N).

The final system of equations is therefore

$$10^4 \begin{bmatrix} 4 & -4 & 0 \\ -4 & 6 & -2 \\ 0 & -2 & 2 \end{bmatrix} \begin{Bmatrix} u_1 \\ u_2 \\ u_3 \end{Bmatrix} = \begin{Bmatrix} 0 \\ 0 \\ 10 \end{Bmatrix} \qquad [2.23]$$

Now constraint conditions must be applied to these equations before solving for the nodal displacements. In this case $u_1 = 0$, and the equations are modified as

$$10^4 \begin{bmatrix} 4 & -4 & 0 \\ -4 & 6 & -2 \\ 0 & -2 & 2 \end{bmatrix} \begin{Bmatrix} 0 \\ u_2 \\ u_3 \end{Bmatrix} = \begin{Bmatrix} 0 + R_1 \\ 0 \\ 10 \end{Bmatrix} \qquad [2.24]$$

where R_1 in the first equation is the reaction provided by the surroundings at node 1 to maintain the zero displacement. Solution of the other two equations yields

$$u_2 = 0.25 \times 10^{-3} \text{ mm}$$
$$u_3 = 0.75 \times 10^{-3} \text{ mm}$$

and substitution back into the first equation of Eq. 2.24 then gives $R_1 = -10$ N as expected (negative because it acts in the negative x direction).

The strains in each element are now calculated using Eq. 2.10, together with the element connectivity information:

$$\varepsilon^{(1)} = (-u_1 + u_2)/L = 0.25 \times 10^{-3}/100 = 2.5 \times 10^{-6}$$
$$\varepsilon^{(2)} = (-u_2 + u_3)/L = 5.0 \times 10^{-6}$$

Finally, the stresses are found by Hooke's law:

$$\sigma^{(1)} = E\varepsilon^{(1)} = 0.5 \text{ N/mm}^2$$
$$\sigma^{(2)} = E\varepsilon^{(2)} = 1.0 \text{ N/mm}^2$$

The theoretical stresses for this problem are easily calculated by

$$\sigma^{(1)}_{\text{theory}} = P/A^{(1)} = 10/20 = 0.5 \text{ N/mm}^2$$
$$\sigma^{(2)}_{\text{theory}} = P/A^{(2)} = 10/10 = 1.0 \text{ N/mm}^2$$

In this example, the finite element analysis has predicted the exact solution to the stepped bar. This is because the variation of displacement through each element was assumed to be linear, and indeed this is the case. Also note that, by selecting a linear distribution, a constant strain and therefore a constant stress field are imposed in each element, which is precisely what is expected in an axially loaded bar of constant cross-sectional area.

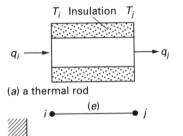

(a) a thermal rod

(b) the equivalent finite element

Figure 2.4 (a) One-dimensional thermal rod (b) equivalent finite element

2.3   A simple one-dimensional element: the thermal rod

The simplest one-dimensional element for heat transfer problems is a thermal rod, as shown in Fig. 2.4(a). The unknown variable at each of the two nodes is temperature T.

Fourier's law states that the heat flow rate q through the rod is proportional to the temperature gradient dT/dx in the direction of conduction. That is,

$$q = -KA \frac{dT}{dx} \qquad [2.25]$$

where q is the heat flow rate, K is the material thermal conductivity and A is the conducting cross-sectional area.

For the finite element of length L in Fig. 2.4(b), if a linear variation of temperature is assumed along the length of the rod then

$$\frac{dT}{dx} = \frac{T_j - T_i}{x_j - x_i} = \frac{T_j - T_i}{L} \qquad [2.26]$$

for calculation of the heat flow from node i. In other words,

$$q_i = -\frac{KA}{L}(T_j - T_i)$$

Similarly it is clear that

$$q_j = -\frac{KA}{L}(T_i - T_j)$$

These two equations can be stored most conveniently in matrix form as

$$\begin{Bmatrix} q_i \\ q_j \end{Bmatrix} = \frac{KA}{L} \begin{bmatrix} 1 & -1 \\ -1 & 1 \end{bmatrix} \begin{Bmatrix} T_i \\ T_j \end{Bmatrix} \qquad\qquad [2.27]$$

This is analogous to Eq. 2.17 with the following substitutions:

$$F \sim q \qquad AE/L \sim KA/L \qquad u \sim T$$

Equation 2.27 is the element equilibrium equation for the one-dimensional thermal rod.

When more than one element is considered, the element stiffness (or conductance) matrices are assembled to give the global stiffness matrix, in the same way as the stiffness matrices of the pin-jointed bar elements were assembled. As a result the system equations are derived, and are of the form

$$\{q\} = [k]\{T\}$$

where $[k]$ is the global stiffness matrix, equal to $\Sigma_{e=1}^{E} [k^{(e)}]$, the sum of the element matrices; $\{q\}$ is the vector of nodal heat flows; and $\{T\}$ is the vector of unknown nodal temperatures.

A simple example now demonstrates the principles.

2.3.1   Heat conduction through a wall

Consider the thermal distribution through the composite wall shown in Fig. 2.5(a), which is composed of a layer of insulation sandwiched between two brick walls.

This can be treated as a one-dimensional problem, and can be modelled with just three elements as in Fig. 2.5(b). The connectivity of the elements is as follows:

Element	i node	j node
(1)	1	2
(2)	2	3
(3)	3	4

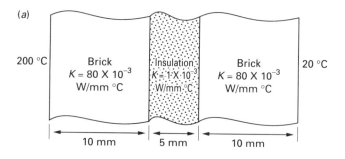

(a)

200 °C

Brick
$K = 80 \times 10^{-3}$
W/mm °C

Insulation
$K = 1 \times 10^{-3}$
W/mm °C

Brick
$K = 80 \times 10^{-3}$
W/mm °C

20 °C

10 mm   5 mm   10 mm

Figure 2.5 Analysis of heat flow through a composite wall: (a) cross-section through wall (b) finite element representation

(b)

1   (1)   2   (2)   3   (3)   4

From Eq. 2.27, assuming a unit cross-sectional area for each element,

$$[k^{(1)}] = \frac{80 \times 10^{-3}}{10}\begin{bmatrix} 1 & -1 \\ -1 & 1 \end{bmatrix} = 8 \times 10^{-3}\begin{bmatrix} 1 & -1 \\ -1 & 1 \end{bmatrix} \text{W/°C}$$

$$[k^{(2)}] = \frac{1 \times 10^{-3}}{5}\begin{bmatrix} 1 & -1 \\ -1 & 1 \end{bmatrix} = 0.2 \times 10^{-3}\begin{bmatrix} 1 & -1 \\ -1 & 1 \end{bmatrix} \text{W/°C}$$

$$[k^{(3)}] = \frac{80 \times 10^{-3}}{10}\begin{bmatrix} 1 & -1 \\ -1 & 1 \end{bmatrix} = 8 \times 10^{-3}\begin{bmatrix} 1 & -1 \\ -1 & 1 \end{bmatrix} \text{W/°C}$$

The global stiffness matrix is found by summing these equations to give

$$[k] = \begin{bmatrix} 8.0 & -8.0 & 0 & 0 \\ -8.0 & 8.0 + 0.2 & -0.2 & 0 \\ 0 & -0.2 & 0.2 + 8.0 & -8.0 \\ 0 & 0 & -8.0 & 8.0 \end{bmatrix} \times 10^{-3} \text{ W/°C}$$

Therefore the full set of system equations with the boundary conditions applied, *i.e.* $T_1 = 200$ °C and $T_4 = 20$ °C, becomes

$$10^{-3}\begin{bmatrix} 8.0 & -8.0 & 0 & 0 \\ -8.0 & 8.2 & -0.2 & 0 \\ 0 & -0.2 & 8.2 & -8.0 \\ 0 & 0 & -8.0 & 8.0 \end{bmatrix}\begin{Bmatrix} 200 \\ T_2 \\ T_3 \\ 20 \end{Bmatrix} = \begin{Bmatrix} Q_1 \\ 0 \\ 0 \\ Q_4 \end{Bmatrix} \qquad [2.28]$$

where Q_1 and Q_4 are the heats applied at nodes 1 and 4 to maintain the temperatures at 200 °C and 20 °C.

Solution of the second and third equations in Eq. 2.28 gives

$$T_2 = 195.71 \text{ °C} \quad \text{and} \quad T_3 = 24.29 \text{ °C}$$

Calculation of the applied heat flows by the first and fourth equations then predicts

$$Q_1 = - Q_4 = 34.32 \times 10^{-3} \text{ watts (per mm}^2)$$

Since the variation of the temperature through the wall thickness is linear, these finite element results agree exactly with those calculated theoretically.

2.4   Conclusions

This chapter introduces two very simple finite elements, and uses each of them to analyse a sample problem. Both the examples are complete, and clearly demonstrate the procedures involved in a typical finite element analysis, namely:

(a)   pre-processing of the model, *i.e.* discretization of the problem into selected element types
(b)   calculation of the element matrices and vectors
(c)   assembly of the element matrices and vectors to give the global (or system) equations
(d)   incorporation of the boundary conditions into the global equations
(e)   solution of the equations to find the unknown nodal values of the field variable
(f)   post-processing of the results to give strains/stresses, heat flows and so on.

Note that each of the sample problems is analysed exactly by the finite element models, but in practice this is most unlikely to occur. It is rare for a finite element model to represent the unknown displacement field precisely, and the results will therefore invariably only approximate the true solution. As the number and complexity of the elements increase, so the approximation should improve and eventually converge to the answer. The skill in finite element modelling comes in developing a sufficiently accurate and representative but economical model of the problem.

Problems

2.1   Prove from first principles that the displacement through a one-dimensional pin-jointed bar element with a linear variation of displacement can be expressed as

$$u = N_i u_i + N_j u_j$$

where $N_i = (x_j - x)/L$ and $N_j = (x - x_i)/L$.

2.2 Using the expanded forms of the strain energy and work done by the loads of a pin-jointed bar (*i.e.* Eqs 2.11 and 2.14), prove that minimization of the potential energy with respect to the nodal displacements u_i and u_j does indeed result in Eq. 2.17.

2.3 Using the calculated values of displacement of the stepped bar in Section 2.2.1, calculate the strain energy in the bar under the 10 N load. Then, by maintaining $u_3 = 0.75 \times 10^{-3}$ mm, vary u_2 and prove that the strain energy is a minimum when $u_2 = 0.25 \times 10^{-3}$ mm.

2.4 What are the displacements of the stepped bar in Section 2.2.1 if a 20 N compressive load is applied at the free end? Calculate the values of the stresses produced, and compare them with the theoretical values. What do you notice about the stiffness matrices for the two load cases?

2.5 Use the finite element method to calculate the displacements and stresses of the bar in Fig. 2.6, and compare with theory.

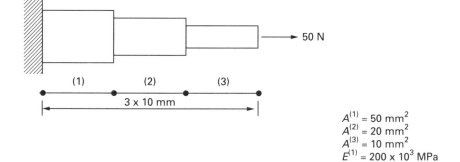

$A^{(1)} = 50 \text{ mm}^2$
$A^{(2)} = 20 \text{ mm}^2$
$A^{(3)} = 10 \text{ mm}^2$
$E^{(1)} = 200 \times 10^3 \text{ MPa}$

Figure 2.6

2.6 What are the temperature and the heat flows of the composite wall of Section 2.3.1 if the temperature on the hotter surface is reduced from 200 °C to 100 °C?

2.7 Use the finite element method to predict the temperature distribution and heat flow through the composite skin shown in Fig. 2.7.

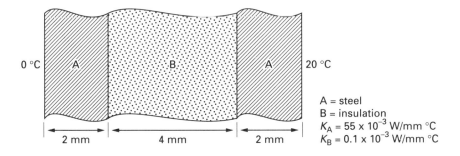

A = steel
B = insulation
$K_A = 55 \times 10^{-3} \text{ W/mm °C}$
$K_B = 0.1 \times 10^{-3} \text{ W/mm °C}$

Figure 2.7

2.8 Ohm's law defines the relationship between the direct current I and the voltage drop $V_2 - V_1$ across a resistor of R ohms. Specifically,

$$I = \frac{1}{R}(V_2 - V_1)$$

By following the method in Section 2.3 for the thermal rod, show that the finite element equations for an electrical element are

$$\begin{Bmatrix} I_1 \\ I_2 \end{Bmatrix} = \frac{1}{R} \begin{bmatrix} 1 & -1 \\ -1 & -1 \end{bmatrix} \begin{Bmatrix} V_1 \\ V_2 \end{Bmatrix}$$

and analyse the circuit in Fig. 2.8.

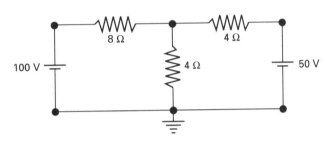

Figure 2.8

2.9 The equation governing the torsion of a circular shaft is

$$T = \frac{JG}{L}\theta$$

where T is the torque, J is the polar second moment of area of the cross-section, G is the shear modulus, L is the length and θ is the twist of the rod (in radians). Derive the finite element equations for a torsion element governed by this equation, and analyse the shaft shown in Fig. 2.9.

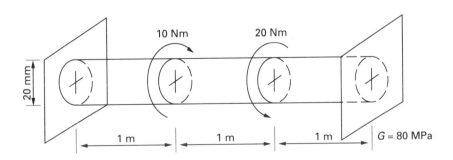

Figure 2.9

3   Discretization of the problem

3.1   Introduction

This is the first step in the finite element procedure, where the body under examination is divided into elements in such a way that the unknown field variable is adequately represented through the body. Care must be taken in the planning and preparation of the problem because considerable savings in time and effort, in both model development and analysis, can be achieved by careful model design. In particular, economies can usually be made by approximating the dimensionality of the problem, and by taking advantage of any symmetry in the body. After considering model simplification, this chapter introduces the range of different elements currently available in commercial finite element packages, and discusses the choice of element type and the size and number of elements that should be used in the model. Following this, the location of the nodes is considered, and the importance of careful node and element numbering is revealed.

3.2   Geometrical approximations

All structures in the real world are three-dimensional. However, just as approximations are made to facilitate simple stress analysis of a part (*i.e.* by assuming plane stress or strain), so the same approach is valid and very useful in finite element modelling. If the geometry and loads of a problem can be completely described in one plane, then the problem can be modelled as two-dimensional. Bodies which are long and whose geometry and loading do not vary significantly in the longitudinal direction can be modelled by using a plane strain representation; for example, consider the analysis of dams and splined shafts (Fig. 3.1). Similarly, bodies that have negligible dimensions in one direction, and are loaded in the plane of the body, can be assumed to be in a plane stress condition, as shown in Fig. 3.2. Heat transfer through a composite wall is examined in the previous chapter, and modelled by using simple one-dimensional elements (Section 2.3). This simplification can be made because it is known that heat flow through the wall is one-dimensional, and, although the situation is rare, the same approach can be used for other problems where all the effects are parallel to or along a straight line.

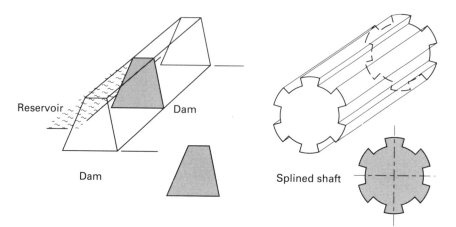

Figure 3.1 Two-dimensional plane strain representations of three-dimensional problems

Reservoir   Dam

Dam   Splined shaft

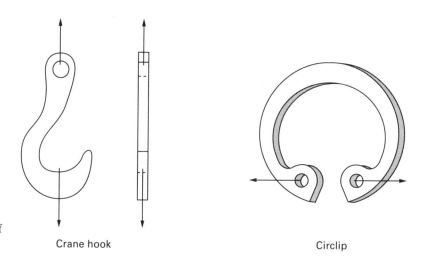

Figure 3.2 Two-dimensional plane stress representations of three-dimensional problems

Crane hook   Circlip

Where a geometrical approximation is not possible, then a full three-dimensional model of the structure will need to be developed, although it might still be possible to limit the size of the model by taking advantage of any symmetry that the problem exhibits.

3.3   Simplification through symmetry

There are four common types of symmetry encountered in engineering problems: axial, planar, cyclic and repetitive, as illustrated in Fig. 3.3. If the configuration of the body and the external conditions (*i.e.* boundary conditions) are similarly symmetric, then only the repeated part of the structure needs to be modelled. It is vital, however, that the loading and constraint conditions are applied to the parts model in such a way that they truly reflect the symmetry of the problem.

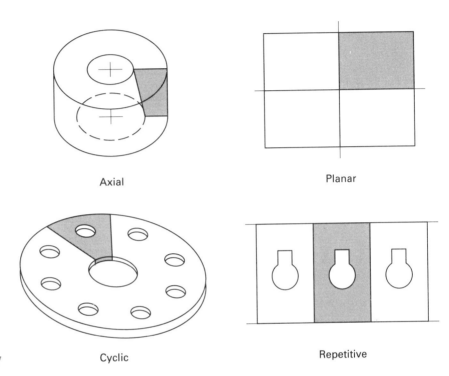

Axial

Planar

Cyclic

Repetitive

Figure 3.3 Types of symmetry

Axial symmetry

Since axial symmetry is encountered so frequently, axisymmetric elements are included in finite element packages. They take account of the constant variable distribution in the circumferential direction, in the same way for example that axisymmetric stress problems are analysed theoretically. This type of problem is clearly similar to those of plane stress and strain, since the distributions and loadings are confined to only two directions (radial and axial). Axisymmetric elements are developed for both stress and thermal analyses in later chapters.

Planar symmetry

Planar symmetry is well illustrated by the case of a flat plate with a hole in it, loaded uniformly as shown in Fig. 3.4. It is only necessary to consider one quarter of the problem, provided the correct constraint conditions are applied to the model. For example, if the deflections in the

x and y directions are u and v respectively, then for this problem u must equal zero along the vertical line of symmetry, and v must equal zero along the horizontal line, as shown.

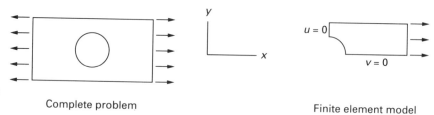

Figure 3.4 Example of problem with planar symmetry

Complete problem

Finite element model

The importance of applying the correct constraint conditions cannot be over emphasized. If they are not modelled precisely, the model will represent a completely different problem.

Cyclic symmetry

Cyclic symmetry is present in spline fittings and propellers, for example (Fig. 3.5). The problems are similar to those described with planar symmetry, except that they are described in a cylindrical rather than a rectangular coordinate system. Naturally the constraint conditions need to be applied in the appropriate direction, *i.e.* the displacements are zero in the circumferential direction for the stress problems.

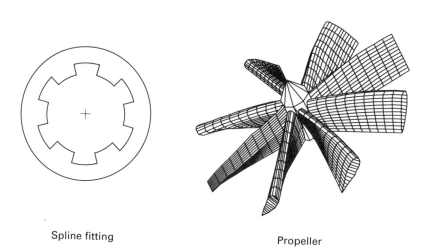

Figure 3.5 Examples of problems showing cyclic symmetry

Spline fitting

Propeller

Repetitive symmetry

Problems exhibiting repetitive symmetry are the least common, and are similar to those structures modelled assuming a plane strain system, as described in Section 3.2. For repetitive symmetry problems the common boundaries of the repeated segment are constrained in a perpendicular direction, as illustrated in Fig. 3.6.

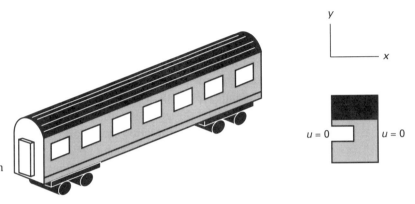

Figure 3.6 Example of problem with repetitive symmetry

3.4   Basic element shapes and behaviour

The basic shapes available in the finite element method are shown in Fig. 3.7. As might be expected, they range from a single point (zero dimensions) through to volumes or three-dimensional shapes.

Shape	Type	Geometry
Point	Mass	
Line	Spring, beam, spar, gap	
Area	2D solid, axisymmetric solid, plate	
Curved area	Shell	
Volume	3D solid	

Figure 3.7 Basic element shapes

Shell elements are in a special category of their own, because they do not neatly fall into either the area or the volume divisions. They are essentially two-dimensional in nature, but are developed so they can be used to model curved surfaces. For stress analysis problems, shell elements are curved plate elements which include both bending and membrane or stretching effects, and are suitable for the modelling of problems such as pressure vessels.

The sides of elements can be straight or curved (Fig. 3.8). If curved elements are used in a model, the time and complexity of the solution are increased significantly for the following reason. The basic finite element method works by approximating the variation of a field variable through each element with a known (interpolation) function, usually a polynomial. However, it is also possible, and necessary with curved elements, to describe the geometry of the chosen element with a polynomial. If the geometric interpolation function and the displacement interpolation function are of the same order, then the element is known as isoparametric, and the two functions prove to be similar to each other, which simplifies their application significantly (this is discussed in greater detail in Chapter 8). Consequently, where possible, isoparametric implementations are used in finite element programs. This means that using elements with curved boundaries implies not only an improvement in the geometric accuracy of the model, but also an increase in the order of the interpolation function used in the element, thus leading to a significant increase in the complexity of the model.

Figure 3.8 Finite elements with curved boundaries

For stress problems, general element behaviour can be classified into one of the following categories:

Membrane   Only in-plane loads are represented, with no bending stiffness normal to the plane.

Bending   Only bending loads are supported.

Plate/shell   A combination of membrane and bending behaviour is used.

Solid   A full three-dimensional stressing regime is available.

Axisymmetric   A constant variable distribution in the circumferential direction is assumed.

When the finite element method is applied to field problems, and in particular thermal problems, the behaviour of the majority of the elements can again be placed into a small number of categories:

Thin shell   A constant temperature is assumed through the element thickness.

Solid   A full three-dimensional thermal field is allowed.

Axisymmetric   The temperature is assumed constant in the circumferential direction.

The behaviour of elements developed for thermal problems is generally much simpler than that of elements used in stress analysis problems, because temperature is a scalar quantity while displacement is a vector.

3.5   Choice of element type

The largest commercial finite element packages, which have facilities to solve stress and a variety of field problems, might easily have more than one hundred different finite elements available for the user. The selection of which element to use in a given problem is, however, not as difficult as it might first appear. Firstly, the **type** of problem to be analysed (*i.e.* stress, field *etc.*) precludes a large number of elements; secondly, the chosen dimensionality of the model restricts the range further. Hence if the user is developing a two-dimensional model of a field problem, there will probably only be one or two suitable element types. Figure 3.9 shows a typical range of elements that a user might expect to find in a commercial program for stress analysis, while Fig. 3.10 shows the type of elements that would normally be offered for thermal or other field problems.

Before choosing the element type, the engineer should try to predict what is taking place in the problem to be examined. Understanding of the basic engineering principles is vital at this stage, and identification of a standard analysis method is very helpful. For example, a component might be recognized as a truss problem, a beam problem or a plate bending problem. In these cases the choice of element is clear.

There is a great temptation for the new user of the finite element method to overmodel a given problem by the use of unnecessarily complex elements. Consider for example the analysis of a simple thin walled cylinder experiencing an internal pressure. There are several ways in which the problem could be modelled, as shown in Fig. 3.11. The simplest of the elements available is the axisymmetric thin shell element, while the most sophisticated is the general three-dimensional solid element. Shell elements (like beam and plate elements) can be formulated in two ways, depending on whether they include the effects of

Element type	Degrees of freedom	Shape
Mass	–	
2D bar *	u, v	
2D beam †	u, v, θ_z	
2D isoparametric solid plane stress plane strain axisymmetric ‡	u, v	
2D interface §	u, v	
Axisymmetric, shell ‡¶	u, v, θ_z	
3D bar *	u, v, w	
3D beam †	u, v, w θ_x, θ_y, θ_z x y	
3D isoparametric solid ‖	u, v, w	
3D shell ¶	u, v, w θ_z, θ_z, θ_z x y	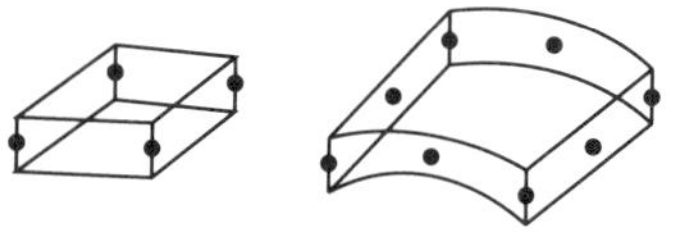
3D interface §	u, v, w	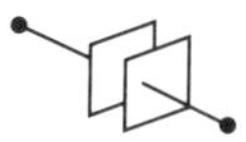

Further possible options:

‡ axisymmetric: including harmonic elements
* bar: tension or compression only
† beam: uniform, tapered, symmetric, unsymmetric
§ interface: gap with/without friction
¶ shell: thick or thin; isotropic or layered
‖ solid: isotropic, anisotropic or layered

Figure 3.9 Typical range of stress elements found in a commercial finite element package

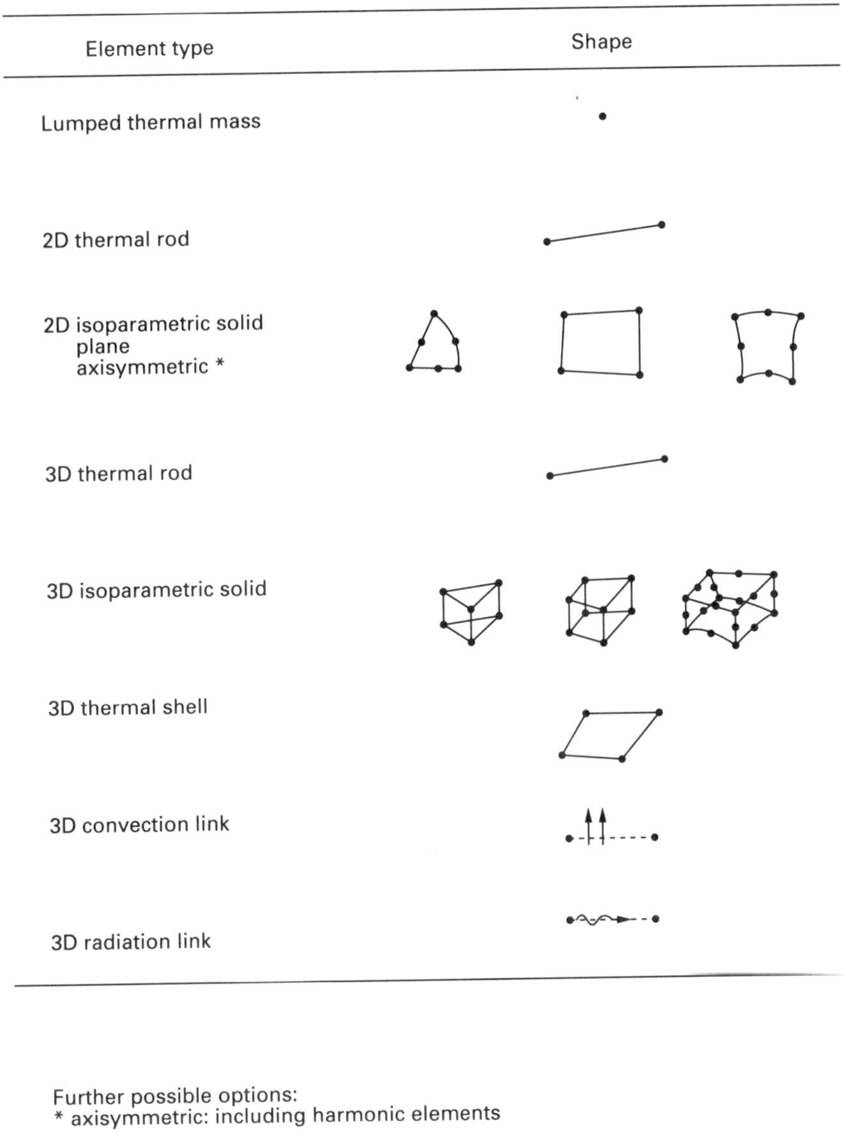

Element type	Shape
Lumped thermal mass	
2D thermal rod	
2D isoparametric solid plane axisymmetric *	
3D thermal rod	
3D isoparametric solid	
3D thermal shell	
3D convection link	
3D radiation link	

Figure 3.10 Typical range of thermal elements found in a commercial finite element package

Further possible options:
* axisymmetric: including harmonic elements

the transverse shear (see Section 8.5). Thin shells ignore transverse shears, and are suitable where the ratio of radius to wall thickness is greater than ten. Thick shells, on the other hand, are used where the thickness and consequently the transverse shear is significant. For the problem in Fig. 3.11, a thin cylinder is being modelled, and consequently a thin shell element formulation should be used. Use of a thick shell

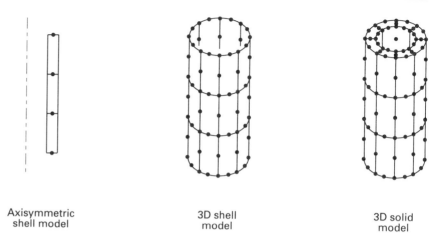

Figure 3.11 Finite element
models of a thin walled cylinder

Axisymmetric
shell model

3D shell
model

3D solid
model

formulation might produce acceptable results but only if the cylinder is
not too thin. Modelling the problem with three-dimensional shell
elements is unnecessary; it will produce no more information than the
axisymmetric model, yet will cost much more. Use of three-dimensional
solid elements will incorrectly estimate the radial stiffness of the cylinder,
lead to ill-conditioning of the equations, and cost significantly more than
the other models. Therefore for this particular problem the axisymmetric
thin shell model is not only the simplest, and the easiest to generate, it is
also the most accurate and, importantly, the cheapest representation. For
example, the cost of an analysis using three-dimensional solid elements
might typically be two or three orders of magnitude more than the
analysis using axisymmetric thin shell elements.

Shell elements are generally available in flat or curved versions. However,

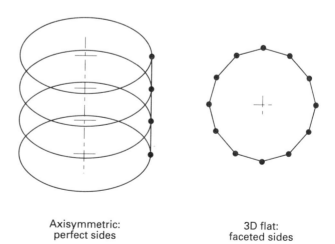

Figure 3.12 Axisymmetric and
three-dimensional flat shell
model of a cylinder

Axisymmetric:
perfect sides

3D flat:
faceted sides

where a finite element program does not have curved shell elements, then such a three-dimensional problem must be modelled with many flat elements to represent the curved geometry, as illustrated in Fig. 3.12. As the number of elements is increased, so the shape approximates a circle, which of course the axisymmetric element models the circle precisely.

3.6   Size and number of elements

The size and number of elements in a finite element model are clearly inversely related. As the number of elements increases, the size of each element must decrease, and consequently the accuracy of the model generally increases, as demonstrated in Fig. 3.13. The problem is the thermal analysis of a cooling fin, where an exact analytical solution shows that the temperature varies in a quadratic manner. If simple one-dimensional elements are used that assume a linear variation in temperature, then the finite element method can be seen to reduce to a procedure of approximating the curved distribution shape with a number of straight lines. The effect of increasing the number of elements is obvious; the finite element solutions gradually approach the true temperature distribution.

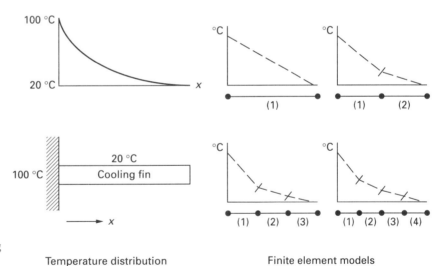

Figure 3.13 Finite element analysis of a cooling fin, showing the effect of increasing element number

Temperature distribution   Finite element models

Sample results of these and other models can be combined into a graph such as Fig. 3.14, which shows how the accuracy of the analysis increases with increasing element numbers. As the number of elements approaches infinity, so the model's prediction will approach the exact solution.

The advantages of using two elements rather than one are clear from

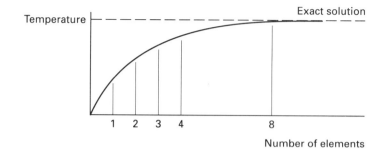

Figure 3.14 Prediction of the
temperature at a point in the
cooling fin

the graph, and four elements produce a quite accurate prediction, but it is
doubtful whether it is worth using eight elements. The increase in
accuracy over the four-element model may only be 5 per cent, and yet the
model size has been doubled. In a commercial environment where every
second of computing time has to be paid for, the importance of this small
increase in accuracy for a doubling of the computing time and cost should
be considered carefully. With the approximations that are inherent in the
majority of engineering calculations, highly accurate results are rarely
worth striving for. Of course in practice it may be difficult to know how
close the results of a finite element model are to the exact solution; hence
every effort should be made to confirm the results of all analyses, even if
it is only possible to confirm the order of the answers. This subject is
discussed in more detail in Section 9.6.

It is assumed in the problem of Fig. 3.13 that the elements are all the
same size, but there is no reason why this should be the case. In fact it is
usual to have many different sized elements in a model. High mesh
density (*i.e.* small elements) should be used where there is an expected
rapid change in the unknown variable. For example, in Fig. 3.13 half of
the rod has a rapidly changing temperature profile, whereas the other half
experiences an almost constant value. Therefore improved accuracy
would have been obtained if the elements had been bunched towards the
hotter end of the rod. Before the engineer starts to mesh the finite
element model, he or she should predict the form of the stress or
temperature distribution that is to be expected, and position the elements
accordingly.

In two and particularly three dimensions, arranging suitable variations
in mesh density usually takes careful consideration and planning. Figure
3.15 shows some of the ways in which the density of simple two-
dimensional meshes of square elements can be varied in a model. Note
that all the nodes must be connected in adjacent elements. The mesh in
Fig. 3.16 is *not* allowed (unless special elements are available); nodes A
and B are not connected to the coarse mesh, which implies that there is a
hole in the material, and the results from such a finite element model
would show a discontinuity at these points.

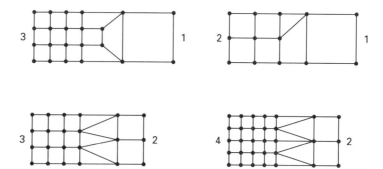

Figure 3.15 Methods of changing mesh density

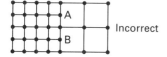

Figure 3.16 Incorrect method of changing mesh density in two-dimensional models

3.7   Element shape and distortion

The finite element method works by approximating the distribution of an unknown variable in a precise manner across the body to be analysed. The process might use linear or quadratic variations through each element. However, these distributions are only reliably produced if the shapes of the elements are not excessively distorted. As element distortion increases, so errors in the element formulations start to become increasingly important. Hence, the elements should all be as 'regular' as possible. For two-dimensional problems, the fundamental element shapes are triangular and rectangular, and the best results are obtained when the first is equilateral and the second square. The allowable limits of distortion are difficult to quantify, and depend very much on the variable distribution that the elements are representing. If the field variable is nearly constant, then even large distortions will not produce significant errors; conversely, rapid variable changes are most sensitive to element shape.

One measure of element distortion is aspect ratio. This is the ratio of the longest side of an element to the shortest side, as illustrated in Fig. 3.17. Other ways of quantifying distortion can be made by measuring the skew and taper of the elements. These allow differentiation of the two rectangular element examples in Fig. 3.17, which have the same aspect ratios but different behaviours. Aspect ratio, skew and taper are considered in more detail in Chapters 9 and 11.

An alternative method of assessing element distortion is to consider the internal angles of the elements. Rectangular elements should include angles as close to 90° as possible, whereas the corners of triangular elements should be near to 60°. As mentioned above, the range of allowable distortion may vary from problem to problem, but as a guide, if no other information is available, the values suggested in Fig. 3.18 may be used.

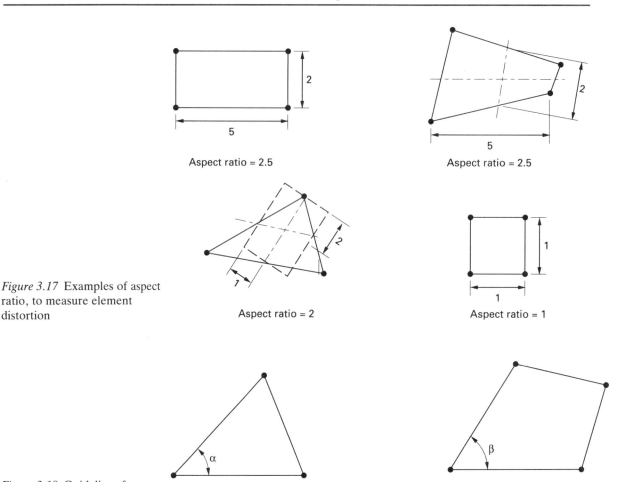

Figure 3.17 Examples of aspect ratio, to measure element distortion

Figure 3.18 Guidelines for allowable element distortions

Some commercial finite element packages perform distortion checking for the user, typically by calculating the aspect ratio and internal angles of the elements. If the values are found to be distorted but within predefined limits, the programs usually print a warning message but continue with the execution of the analysis. However, if the elements are grossly distorted the programs stop, and proceed only with specific authorization from the user.

When the elements have mid-side nodes, then their sides can be curved to follow small curvatures in the problem under analysis. Again however, as the sides are curved, so errors are introduced into the calculations. The 'best' shape is one in which the element sides are flat, and the node is placed equally between the corner nodes. As a general rule, the limits shown in Fig. 3.19 should be followed.

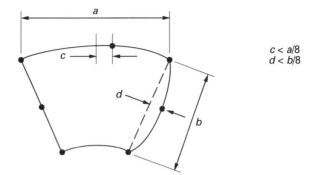

Figure 3.19 Guidelines for curved sides in higher-order elements

3.8   Location of nodes

So far, the discussions in this chapter have implied that, when modelling a problem, the user needs only to concentrate on the division of the geometry into elements so that the variable distribution is adequately represented. However, the location of the elements and therefore of the nodes must also reflect any changes in material properties, geometry, constraint conditions and applied loads. The following examples illustrate this point.

The stiffness or conductance matrix of each element depends on the material properties of that region in the model, and consequently an element cannot comprise two different materials. Therefore a line (or area) of nodes will always be required at the interface of different materials, as in Fig. 3.20(a).

If there is an abrupt change in cross-section of a one-dimensional model, for example, a node would be required at that point. Similarly, a discontinuity in a two-dimensional or three-dimensional model, from a crack possibly, would need a line (or area) of nodes defining the gap in the material, as shown in Fig. 3.20(b).

When a concentrated load is applied in a stress analysis problem, then there must be a node at the corresponding position in the finite element model. Or, if a distributed load is present, then nodes must define the start and finish positions of the load, as in Fig. 3.20(c). The corresponding case in heat transfer problems might be convection or incident heat flux acting on a surface of the body, as in Fig. 3.20(d).

These examples may appear obvious, but in complex modelling situations the careful consideration of discontinuities in material properties, geometry, constraint conditions and applied loads is vital prior to construction of the model, to ensure that the nodes are placed correctly.

(a)

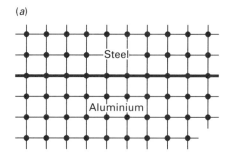

(b)

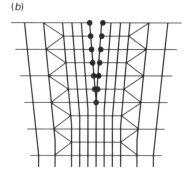

(c)

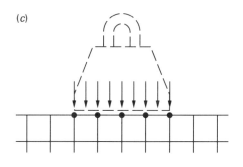

(d)

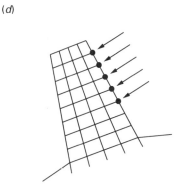

Figure 3.20 Examples of two-dimensional models where node location is important, showing a line of nodes (a) at the interface of different materials (b) around a discontinuity (c) at an area of distributed load (d) defining a length of model experiencing an incident heat flux

3.9   Node and element numbering

Remember that after the equations for each element are derived, they are combined to give the system equations for the whole problem. These are then solved to yield the nodal values of the unknown variable, *i.e.* the displacements or temperatures. There are two methods that are commonly employed to solve the system equations: Gaussian elimination and the wavefront solution technique. The method used depends on the complexity of the program.

Commercial finite element packages, which have the ability to analyse thousands of elements in a single model, tend to use the wavefront method. This is because although it is more complex to implement, the actual computer memory required to store and process the model data is reduced. Gaussian elimination, on the other hand, is straightforward to apply but does require more memory to solve the equations.

Full details of the assembly and solution of the equations are given in Chapter 7, but the solution methods are outlined here because they have a direct bearing on the way the model is constructed. Hence before a model is developed the user must be aware of which method the program uses.

Bandwidth

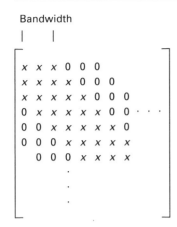

Figure 3.21 Type of banded matrix produced in a finite element analysis

Solution by Gaussian elimination

The simple examples of Sections 2.2.1 and 2.3.1 illustrated how the element equations are assembled into the global matrix. Note particularly that the global matrix in Eq. 2.28 has several zero off-diagonal terms. In fact, the matrix is banded. This is always a feature of finite element models. A band of non-zero terms surrounds the diagonal values (which will always be non-zero), and the remaining locations in the matrix are filled with zeros; this is shown in Fig. 3.21. Notice the bandwidth of the matrix. Obviously the smaller the bandwidth, then the easier and faster the matrix can be solved by an elimination method. It will be shown below that bandwidth depends directly on the order that the nodes are numbered.

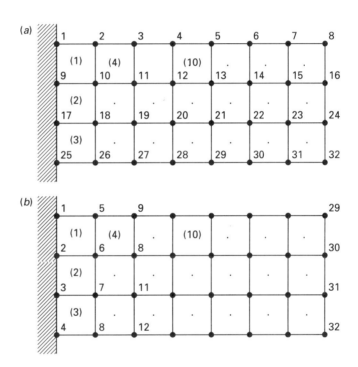

Figure 3.22 Two different methods of labelling the nodes: (a) horizontally (b) vertically

Consider, for example, the two plane stress models of a short beam in Fig. 3.22. The models are the same except for the definitions on the nodes, which are labelled either horizontally or vertically. From the model with horizontal labelling we can see that the degrees of freedom associated with nodes 1 and 10 will be related since they occur in the same element. However, in the model with vertical labelling the highest node that node 1 will be related to is number 6. Now since the models are two-dimensional, each node will have two degrees of freedom; the first two columns and first two rows of the stiffness matrix will correspond to node 1, and the eleventh and twelfth columns and rows will correspond to node 6. Therefore the bandwidth of the vertically labelled model will be 12, while that of the horizontally labelled model will be 20 (columns 1 to 20). The same calculations could be performed for each element in the corresponding models. For example, consider element (10). For Fig. 3.22(a), nodes 4 and 13, or columns 7 and 8 to 25 and 26, are related, again implying a bandwidth of 20. In general, the bandwidth can be calculated by

$$\text{bandwidth} = (D + 1)f$$

where D is the largest difference between nodes in a single element and f is the number of degrees of freedom at each node.

It should be obvious from the above example that the order of node numbering is vital when Gaussian elimination is used; in particular, the nodes should be labelled across the shortest dimension to achieve the smallest nodal difference in any one element.

Solution by the wavefront method

The wavefront method actually never constructs the whole stiffness matrix. Instead it eliminates the degrees of freedom as soon as it can when it works through the model. The elements are analysed one at a time, and their individual stiffness matrices are assembled into a temporary matrix as the solution proceeds. When the last occurrence of a node is noted, its degrees of freedom are eliminated out of the matrix and stored for later calculation.

For example, consider Fig. 3.22(a). The stiffness matrix of element (1) is calculated and placed into the temporary matrix. Since node 1 only occurs in the first element it will not be referenced by any other element and it can be removed from the matrix. This means that the degrees of freedom associated with node 1 are written as functions of the degrees of freedom of nodes 2, 9 and 10. Therefore the stiffness matrix will only contain information on these three nodes. Element (2) is considered next; its stiffness matrix is calculated and added into the temporary matrix. Node 9 is not used by any other element and can be removed; hence, at this point, only information on nodes 2, 10, 17 and 18 is held in the matrix. This process continues through all the elements until only node 32 is left, and its degrees of freedom are solved. Then a backward substitution occurs and all the other degrees of freedom are evaluated, in the same way as the back substitution in Gaussian elimination.

The method is called a wavefront technique because the nodes held in the matrix move like a wave through the model, as illustrated in Fig. 3.23. If the elements had been labelled down the length of the beam, then

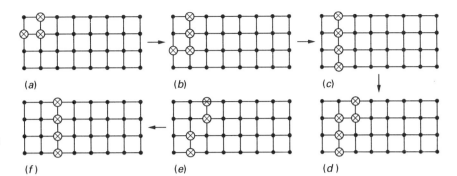

Figure 3.23 Movement of the wave down the length of the model

at any one time there would have been more nodes in the wavefront and consequently more nodes in the temporary matrix, and the solution would have been considerably slower. Clearly, for the wavefront solution method, the numbering of the elements is crucial; they should be ordered across the shortest dimension of the model.

In conclusion, then, smaller programs will probably use Gaussian elimination to solve the system equations, in which case the nodes should be labelled across the shortest dimension of the model. The element numbering in these analyses is immaterial. However, for large commercial packages the wavefront method will be used, and the element numbering is vital. The elements should be labelled across the shortest dimension, but the labelling of the nodes is of no importance.

3.10   Conclusions

This chapter introduces the first basic steps in finite element modelling. Some of the topics will be expanded later in this book. There has been no attempt to describe exactly how the model is defined when a finite element package is being used, since that depends very much on the program itself, and varies from one program to another. However, the general facilities (but not the specific commands) that can be expected from currently available commercial packages are described in Chapter 12. The new user will probably be surprised at how sophisticated some of the model generating facilities are. The engineer must still make the important decisions, such as type and number of elements, but the programs do much (if not all) of the tedious calculations, and the data entry is reduced to a minimum. This level of sophistication does encourage overmodelling of a problem, since it is relatively easy to develop the model, and one command might be all that is necessary to mesh 1000 elements instead of 100. Careful planning of the model is vital. It is bad practice to use a three-dimensional model where a two-dimensional or axisymmetric model would be adequate, and unnecessarily complex elements should not be used unless the problem really demands it.

Problems

3.1 What (if any) geometrical and/or symmetrical approximations can be made in the following finite element investigations, assuming representative loading and constraint conditions:

(a)   stress analysis of a tensile test specimen
(b)   stress analysis of a G-clamp
(c)   stress analysis of a car wheel
(d)   stress analysis of a bicycle frame
(e)   stress analysis of a tooth
(f)   stress analysis of a domestic hot water cylinder
(g)   stress analysis of a gas turbine blade
(h)   stress analysis of a suspension bridge
(i)   stress analysis of a spur gear
(j)   stress analysis of a motor car's crankshaft
(k)   thermal analysis of a central heating radiator
(l)   thermal analysis of a gas turbine blade
(m)   thermal analysis of a motor car's piston
(n)   acoustic analysis of an aircraft cabin
(o)   flow analysis round a long cylinder
(p)   flow analysis round a motor car in a wind tunnel
(q)   flow analysis under a dam
(r)   frequency analysis of a square clamped plate
(s)   frequency analysis of a tuning fork
(t)   frequency analysis of a helicopter rotor blade?

3.2 Consider what type of element(s) might be used in the stress and thermal analyses listed in Problem 3.1.

4 Interpolation functions and simplex elements

4.1 Introduction

The finite element method works by assuming a given distribution of the unknown variable(s) through each element. In the examples of Sections 2.2.1 and 2.3.1, simple one-dimensional elements are introduced which use a linear variation of displacement or temperature. The equations defining the approximating distribution are known as interpolation functions, and can take any mathematical form, although in practice they are usually polynomials. The linear interpolation functions already introduced are of course polynomials of the first order.

Polynomials are popular because they are easy to formulate and computerize, and in particular their differentiation and integration is straightforward to implement on a computer. With polynomial functions, the accuracy of the analysis can also be simply improved by increasing the order of the polynomial, as demonstrated in Fig. 4.1. The graphs show the variation of temperature across a region or element of a body. The solid line represents the actual (unknown) distribution, and the other lines show constant, linear and quadratic approximations to the problem. Clearly the higher the order, the more closely the real variation is modelled, so that an element employing the quadratic interpolation function would produce more accurate results than an element using one of the other functions.

Of course there is a price to pay for the increased accuracy. As the order of the interpolation function increases, so do the number of calculations necessary to evaluate the model. Typically, a quadratic element will require two or three times the amount of computer time that a linear element would need, although, as the number of dimensions of the problem increases, so the differences become even more significant. In some problems, therefore, it may be more economical to use a finer mesh of linear elements. To set up an efficient model of a problem, it is

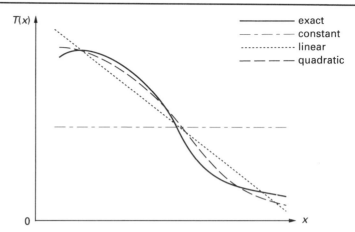

Figure 4.1 Polynominal approximation in one dimension

clearly desirable if the engineer can predict the form of the stress or thermal distribution, and arrange the elements accordingly. This subject will be discussed again many times in the following chapters.

For the one-dimensional thermal rod, a linear interpolation function (as used in Section 2.3.1) takes the form

$$T = \alpha_1 + \alpha_2 x$$

where α_1 and α_2 are constants, and are evaluated by substituting $T = T_i$ and $T = T_j$ at $x = x_i$ and $x = x_j$ respectively. The quadratic form of the interpolation function will be

$$T = \alpha_1 + \alpha_2 x + \alpha_3 x^2$$

and a cubic function would be

$$T = \alpha_1 + \alpha_2 x + \alpha_3 x^2 + \alpha_4 x^3$$

Since the constants are calculated from the geometry of the element over which the function acts, as the order and number of constants increase so it is to be expected that the number of nodes defining the element increases. Therefore an element with a cubic interpolation function (and four constants) would require four nodes, as shown in Fig. 4.2.

Interpolation function	Nodes	Element
$T = \alpha_1 + \alpha_2 x$	2	•————————•
$T = \alpha_1 + \alpha_2 x + \alpha_3 x^2$	3	•————•————•
$T = \alpha_1 + \alpha_2 x + \alpha_3 x^2 + \alpha_4 x^3$	4	•——•——•——•

Figure 4.2 One-dimensional elements and their interpolation functions

The extension of this idea to two and three dimensions is easy. A linear variation of temperature across a two-dimensional element, for example, would be defined by the function

$$T = \alpha_1 + \alpha_2 x + \alpha_3 y$$

This has three unknown constants, so the element must have three nodes. In two dimensions this means the element has a triangular shape. A higher-order element is constructed by adding more terms to the interpolation function. For a quadratic approximation in two dimensions,

$$T = \alpha_1 + \alpha_2 x + \alpha_3 y + \alpha_4 xy + \alpha_5 x^2 + \alpha_6 y^2$$

which includes all the second-order terms. In the same way that the higher-order one-dimensional elements are developed by introducing extra nodes between the two end nodes of the original one-dimensional element, so the two-dimensional family of elements takes the form presented in Fig. 4.3.

Interpolation function	Nodes	Element
$T = \alpha_1 + \alpha_2 x + \alpha_3 y$	3	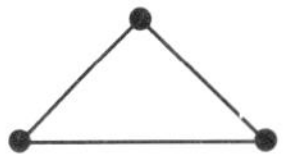
$T = \alpha_1 + \alpha_2 x + \alpha_3 y + \alpha_4 xy + \alpha_5 x^2 + \alpha_6 y^2$	6	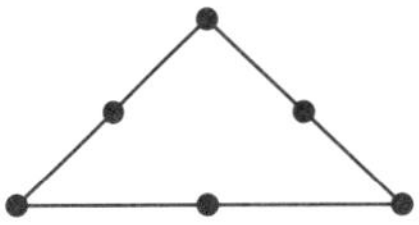

Figure 4.3 Two-dimensional elements and their interpolation functions

Note that there is no requirement to use pure linear, quadratic or cubic types of function. The (two-dimensional) polynomial

$$T = \alpha_1 + \alpha_2 x + \alpha_3 y + \alpha_4 xy$$

is a very popular interpolation equation. It has four unknown constants with four nodes defining a rectangular element, and it is discussed in the following section.

4.2   Simplex, complex and multiplex elements

Elements are normally categorized by virtue of their interpolation functions into three groups: simplex, complex and multiplex. Typical elements for the different categories are shown in Fig. 4.4.

Simplex elements have polynomials with constant and linear terms only, and the nodes are located at the corners of the elements. For

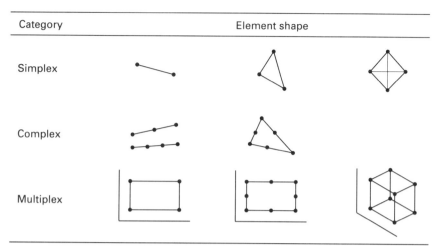

Category	Element shape
Simplex	
Complex	
Multiplex	

Figure 4.4 Examples of simplex, complex and multiplex elements

example, the two-dimensional simplex element is a triangle with three corner nodes. The three-dimensional simplex element is a tetrahedron with four corner nodes.

Complex elements use quadratic, cubic and higher-order interpolation polynomials. They have the same shape as the simplex elements but have additional nodes, usually on the boundaries of the elements but although sometimes internally.

Multiplex elements also use high-order interpolation polynomials, but unlike complex elements do not use complete polynomial expressions. To ensure continuity between different elements, the sides of multiplex elements have to be parallel to the coordinate system. For example, the interpolation function

$$T = a_1 + a_2 x + a_3 y + a_4 xy$$

which has one more term than the usual two-dimensional simplex element, defines the variation of T over a rectangle with sides parallel to the x and y axes.

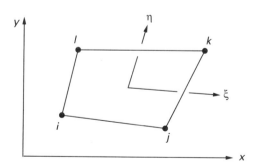

Figure 4.5 Multiplex element in a curvilinear coordinate system

At first sight it appears that multiplex elements will only be of limited use, because of their requirement to have sides parallel to the coordinate axes. However, this can be overcome by using a curvilinear coordinate system to define each element, as shown in Fig. 4.5, where the variation of T is defined in the coordinate system (ξ, η). By this technique, multiplex elements turn out to be the most widely used type of element. They are discussed in detail in Chapter 8.

4.3   Linear interpolation polynomials for simplex elements

The previous section shows that a wide range of elements is possible, and this chapter begins by discussing simplex elements, leaving the higher-order elements until later in the book. The finite element equations relating to simplex elements are the easiest to handle, and in particular can be integrated exactly. Higher-order elements, on the other hand, require some form of numerical integration, which although not difficult does add further steps to the procedures.

In the following pages, one-, two- and three-dimensional simplex elements are introduced in which each element is identified by general node labels such as i, j and k. In modelling, these labels are substituted by the real node numbers. The location of the first and last node is immaterial, provided that the elements are labelled consistently – for example, in an anticlockwise direction for the two-dimensional triangular element.

The one-dimensional simplex element

This element is used in the simple examples of Sections 2.2.1 and 2.3.1. The general form of the interpolation function is

$$\phi = \alpha_1 + \alpha_2 x \qquad\qquad [4.1]$$

which defines the distribution of the unknown variable (possibly displacement or temperature) in the one-dimensional element of length L, as shown in Fig. 4.6. The element has two nodes i and j, which are placed at distances x_i and x_j from the origin. By substituting the value of ϕ at each node into Eq. 4.1 it is found that

$$\alpha_1 = \frac{\phi_i x_j - \phi_j x_i}{x_j - x_i} = \frac{\phi_i x_j - \phi_j x_i}{L}$$

$$\alpha_2 = \frac{\phi_j - \phi_i}{x_j - x_i} = \frac{\phi_j - \phi_i}{L}$$

and substituting these back into Eq. 4.1 gives

$$\phi = \frac{\phi_i x_j - \phi_j x_i}{L} + \frac{\phi_j - \phi_i}{L} x \qquad [4.2]$$

Equation 4.2 can be rearranged in terms of ϕ_i and ϕ_j as

$$\phi = \frac{x_j - x}{L} \phi_i + \frac{x - x_i}{L} \phi_j \qquad [4.3]$$

or

$$\phi = N_i \phi_i + N_j \phi_j = [N]\{\Phi\} \qquad [4.4]$$

where N_i and N_j are called shape functions. Each shape function is associated with one particular node, identified by the subscript. It is a property of a shape function that it equals unity at its own assigned node and zero at the other nodes of the element. This is easily confirmed for the one-dimensional element by considering what happens to N_i and N_j at $x = x_i$ and $x = x_j$. Also the sum of all the shape functions at any point in an element equals one. For convenience the shape functions are stored in a shape function matrix $[N]$, and the nodal variables are stored in a vector $\{\Phi\}$.

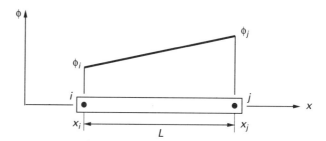

Figure 4.6 One-dimensional simplex element

Note that the shape functions, like the original interpolation polynomial, are linear functions of x. It is always the case that the shape functions are polynomials of the same order as the interpolation function. Therefore shape functions N_i and N_j of a one-dimensional simplex element have the distribution shown in Fig. 4.7(a). For a one-dimensional element with a quadratic interpolation function and three nodes, the shape functions take the quadratic form shown in Fig. 4.7(b). As expected, the values of the three shape functions vary between zero and unity.

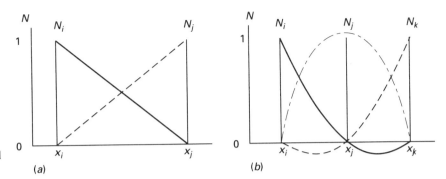

Figure 4.7 Variation of the shape function in (a) linear and (b) quadratic elements

The two-dimensional simplex element

This is a straight sided triangle with three nodes, with an interpolation function

$$\phi = \alpha_1 + \alpha_2 x + \alpha_3 y \qquad [4.5]$$

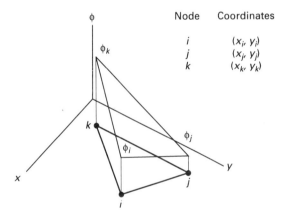

Node	Coordinates
i	(x_i, y_i)
j	(x_j, y_j)
k	(x_k, y_k)

Figure 4.8 Two-dimensional simplex element

If the values of ϕ at each node are ϕ_i, ϕ_j and ϕ_k as shown in Fig. 4.8, the values of the α constants are found to be

$$\alpha_1 = (a_i\phi_i + a_j\phi_j + a_k\phi_k)/2A$$
$$\alpha_2 = (b_i\phi_i + b_j\phi_j + b_k\phi_k)/2A \qquad [4.6]$$
$$\alpha_3 = (c_i\phi_i + c_j\phi_j + c_k\phi_k)/2A$$

where A is the area of the triangle, given by

$$A = \frac{1}{2}\begin{vmatrix} 1 & x_i & y_i \\ 1 & x_j & y_j \\ 1 & x_k & y_k \end{vmatrix} = \frac{1}{2}(x_iy_j + x_jy_k + x_ky_i - x_iy_k - x_jy_i - x_ky_j)$$

$$[4.7]$$

and where

$$
\begin{aligned}
a_i &= x_j y_k - x_k y_j & b_i &= y_j - y_k & c_i &= x_k - x_j \\
a_j &= x_k y_i - x_i y_k & b_j &= y_k - y_i & c_j &= x_i - x_k \\
a_k &= x_i y_j - x_j y_i & b_k &= y_i - y_j & c_k &= x_j - x_i
\end{aligned} \tag{4.8}
$$

Using Eqs 4.5 and 4.6, the interpolation function can be reformatted to

$$
\phi = N_i \phi_i + N_j \phi_j + N_k \phi_k = [N]\{\Phi\} \tag{4.9}
$$

where

$$
\begin{aligned}
N_i &= (a_i + b_i x + c_i y)/2A \\
N_j &= (a_j + b_j x + c_j y)/2A \\
N_k &= (a_k + b_k x + c_k y)/2A
\end{aligned} \tag{4.10}
$$

In other words, the shape functions of a two-dimensional simplex element have the general form

$$
N_\beta = (a_\beta + b_\beta x + c_\beta y)/2A \qquad \beta = i,j,k \tag{4.11}
$$

As before, each shape function equals unity at one node, and zero at the other two. For the linear functions in Eq. 4.10, this results in lines of constant value across the element. For example, the variation of N_j is shown in Fig. 4.9.

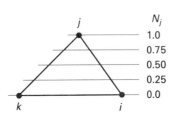

Figure 4.9 Variation of the shape function N_j across the two-dimensional simplex element

Example 4.1: calculation of shape functions

Calculate the shape functions of all the nodes in element (3) and that of node 2 in element (4) of the mesh shown in Fig. 4.10.

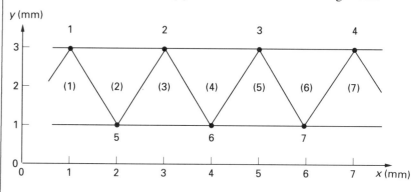

Figure 4.10

Remember that the equations derived above are for a general element. The nodes are labelled i, j and k, but these need to be replaced with the true node numbers for the specific elements under consideration. It is immaterial which node in an element is selected as node i, but the nodes in the elements must all be labelled in a systematic manner, e.g. all anticlockwise.

For elements (3) and (4) of this particular problem, let the mode numbering be as follows:

Element	i node	j node	k node
(3)	6	2	5
(4)	3	2	6

The coordinates of these nodal points are summarized as follows:

Coordinate	Node			
	2	3	5	6
x	3	5	2	4
y	3	3	1	1

Therefore, for element (3),

$$a_6 = 3 \times 1 - 2 \times 3 = -3 \quad b_6 = 3 - 1 = 2 \quad c_6 = 2 - 3 = -1$$
$$a_2 = 2 \times 1 - 4 \times 1 = -2 \quad b_2 = 1 - 1 = 0 \quad c_2 = 4 - 2 = 2$$
$$a_5 = 4 \times 3 - 3 \times 1 = -9 \quad b_5 = 1 - 3 = -2 \quad c_5 = 3 - 4 = -1$$

The area of the element can be calculated by Eq. 4.7, or more easily by $(0.5 \times \text{base} \times \text{height}) = 2 \text{ mm}^2$.

The shape functions for element (3) are then found from Eq. 4.10 to be

$$N_6^{(3)} = 0.25 \, (-3 + 2x - y)$$
$$N_2^{(3)} = 0.25 \, (-2 + 2y) \qquad\qquad [4.12]$$
$$N_5^{(3)} = 0.25 \, (9 - 2x - y)$$

Note that a superscript has been added to identify the fact that these are the shape functions of element (3).

Considering the shape function of node 6, then substitution of the coordinates of the three nodes in the element gives

$$N_6^{(3)} = 1 \quad \text{at node 6}$$
$$N_6^{(3)} = 0 \quad \text{at nodes 2 and 5}$$

as expected.

For node 2 in element (4),

$$a_2 = 4 \times 3 - 5 \times 1 = 7 \quad b_2 = 1 - 3 = -2 \quad c_2 = 5 - 4 = 1$$

and the area also equals 2 mm^2, so that

$$N_2^{(4)} = 0.25 \, (7 - 2x + y)$$

Note that this is entirely different to the shape function for node 2 of element (3), given in Eq. 4.12. The shape functions are individual to each element, since they rely solely on the geometry of the particular element.

The three-dimensional simplex element

The principles discussed so far can be directly extended to three dimensions. The basic three-dimensional simplex element is a flat faced tetrahedron with four nodes, such that i, j and k are labelled anticlockwise on any face as viewed from the vertex opposite the face, which is labelled l (Fig. 4.11). The general interpolation polynomial for the element is

$$\phi = \alpha_1 + \alpha_2 x + \alpha_3 y + \alpha_4 z \qquad [4.13]$$

which can be expressed in terms of the nodal values of ϕ and the nodal shape functions as

$$\phi = N_i \phi_i + N_j \phi_j + N_k \phi_k + N_l \phi_l = [N]\{\Phi\} \qquad [4.14]$$

where the shape functions are given by

$$N_\beta = (a_\beta + b_\beta x + c_\beta y + d_\beta z)/6V \qquad \beta = i,j,k,l \qquad [4.15]$$

The a, b, c and d constants are functions of the nodal coordinates, and generally take the form

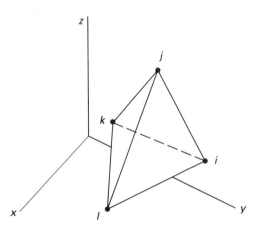

Figure 4.11 Three-dimensional simplex element

$$a_i = \begin{vmatrix} x_j & y_j & z_j \\ x_k & y_k & z_k \\ x_l & y_l & z_l \end{vmatrix} \qquad b_i = \begin{vmatrix} 1 & y_j & z_j \\ 1 & y_k & z_k \\ 1 & y_l & z_l \end{vmatrix}$$

$$[4.16]$$

$$c_i = \begin{vmatrix} x_j & 1 & z_j \\ x_k & 1 & z_k \\ x_l & 1 & z_l \end{vmatrix} \qquad d_i = \begin{vmatrix} x_j & y_j & 1 \\ x_k & y_k & 1 \\ x_l & y_l & 1 \end{vmatrix}$$

The coefficients in these expressions form a pattern, and the constants relating to the other nodes are found by cyclic interchange of the subscripts in the order i, j, k, l.

4.4   Natural coordinates

When the governing element equations are derived in the next chapters, it will be found that many terms need to be calculated by the integration of some function of the shape functions. This is not difficult, since the shape functions are simple linear terms of x, y and z, but it can be made even easier by the introduction of natural coordinates. These are coordinate systems that are local (*i.e.* individual) to each element, but are dimensionless and have a maximum absolute magnitude of unity.

One-dimensional natural coordinates

For a one-dimensional element a point P is identified by two natural coordinates L_1 and L_2, which are defined by the ratios

$$L_1 = \frac{l_1}{L} = \frac{x_j - x}{L}$$

$$L_2 = \frac{l_2}{L} = \frac{x - x_i}{L}$$

[4.17]

where l_1 and l_2 are simply the distances from the point to the nodes, as shown in Fig. 4.12. Looking back to Eq. 4.3, it is clear that the natural coordinates are directly equivalent to the shape functions for the element, so that

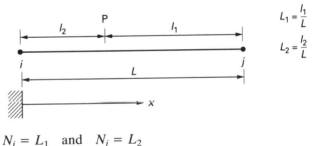

Figure 4.12 Natural coordinates of a one-dimensional element

$$N_i = L_1 \quad \text{and} \quad N_j = L_2$$

[4.18]

Instead of the variation of ϕ through the element, the natural coordinates can be used to describe the geometry of the element. This is of little value for a one-dimensional element, but for two- and three-dimensional elements and particularly higher-order elements it is a very important concept. In one dimension

$$x = L_1 x_i + L_2 x_j = N_i x_i + N_j x_j$$

[4.19]

which is comparable to Eq. 4.4.

It has been proved that the integration over the element of functions incorporating L_1 and L_2 raised to the powers α and β can be easily calculated by applying the formula

$$\int_L L_1^\alpha L_2^\beta \, dx = \frac{\alpha! \beta!}{(\alpha + \beta + 1)!} L \qquad [4.20]$$

where $\alpha!$ is factorial α; for example, $3! = 3 \times 2 \times 1 = 6$.

Therefore, since $L_1 = N_i$ and $L_2 = N_j$, the same equation can also be applied where integrations involving the shape functions are required.

Example 4.2: integration formulae

If $x_i = 2$ and $x_j = 6$, evaluate the following integral by long-hand and by Eq. 4.20:

$$\int_L L_1^2 L_2 \, dx$$

The natural coordinates are $L_1 = (6 - x)/4$ and $L_2 = (x - 2)/4$, so that

$$\int_2^6 L_1^2 L_2 \, dx = \frac{1}{64} \int_2^6 (6 - x)^2 (x - 2) \, dx$$

$$= \frac{1}{64} \int_2^6 (x^3 - 14x^2 + 60x - 72) \, dx$$

$$= \frac{1}{64} \left[\frac{x^4}{4} - \frac{14x^3}{3} + 30x^2 - 72x \right]_2^6 = 0.333$$

or

$$\int_2^6 L_1^2 L_2 \, dx = \frac{2!1!}{(2 + 1 + 1)!} L = \frac{2 \times 1 \times 1}{4 \times 3 \times 2 \times 1} \times 4 = 0.333$$

Two-dimensional local coordinates

When the concept of natural coordinates is applied to two dimensions, the result is an area (or triangular) coordinate. For example, consider Fig. 4.13 which shows the three coordinates L_1, L_2 and L_3 defining the point P. The natural coordinates are defined by the ratios of the areas opposite each node and the total area of the triangular element:

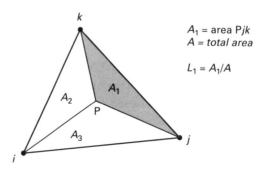

A_1 = area Pjk
A = total area

$L_1 = A_1/A$

Figure 4.13 Calculation of the natural coordinates of a two-dimensional element

$$L_1 = A_1/A \qquad L_2 = A_2/A \qquad L_3 = A_3/A \qquad [4.21]$$

Examination of Fig. 4.13 shows that the three area coordinates are related since

$$A_1 + A_2 + A_3 = A$$

or

$$\frac{A_1}{A} + \frac{A_2}{A} + \frac{A_3}{A} = L_1 + L_2 + L_3 = 1 \qquad [4.22]$$

Also, consideration of A_1 in Fig. 4.13 shows that it will vary from 0 to A as point P moves from side jk to node i. In other words, the coordinate L_1 varies from 0 to 1. This is precisely the same way that the shape function corresponding to node i varied (Fig. 4.9). Not surprisingly it is found that

$$L_1 = N_i \qquad L_2 = N_j \qquad L_3 = N_k \qquad [4.23]$$

If the natural coordinates of a point P are known, it is now possible to calculate the global coordinates (x, y) of the point from the following equations, which contain the global coordinates of the nodes and the natural coordinates of the point P:

$$x = L_1 x_i + L_2 x_j + L_3 x_k$$
$$y = L_1 y_i + L_2 y_j + L_3 y_k \qquad [4.24]$$

Example 4.3: conversion from natural to global coordinates

Calculate the global coordinates of a point in element (3) of the mesh in Fig. 4.10, which is defined by $L_1 = L_2 = 1/3$ using the same nodal order as in Example 4.1.

To find the global coordinates of a point with the given natural coordinates, the first step is to calculate the third unspecified coordinate L_3. Using Eq. 4.22, $L_3 = 1/3$, since the sum of L_1, L_2 and L_3 must equal unity.

Hence, using Eqs 4.24 and the known nodal coordinates,

$$x = \frac{1}{3}(4 + 3 + 2) = 3.0 \text{ mm}$$

$$y = \frac{1}{3}(1 + 3 + 1) = 1.67 \text{ mm}$$

With the given values of the natural coordinates, it should be obvious that point P is in fact the centroid of the element, and examination of Fig. 4.10 shows that the centroid of element (3) is indeed at (3, 1.67).

As with the one-dimensional example in the previous section, formulae have been developed which allow easy integration of functions containing the natural coordinates, and therefore similarly easy evaluation of functions of the shape functions. The integration formulae are

$$\int_H L_1^\alpha L_2^\beta \, dH = \frac{\alpha! \beta!}{(\alpha + \beta + 1)!} H \qquad [4.25]$$

where H is the distance between two nodes on a side of the element, and

$$\int_A L_1^\alpha L_2^\beta L_3^\gamma \, dA = \frac{\alpha! \beta! \gamma!}{(\alpha + \beta + \gamma + 2)!} 2A \qquad [4.26]$$

where A is the face area of the element.

Hence, in the derivation of the element equations of a stress or thermal problem, where it is necessary to integrate terms containing the shape functions, these equations may be applied to simplify the calculations.

Example 4.4: calculation of a stiffness matrix using integration formulae

The derivation of the stiffness matrix of two-dimensional field problems involves the calculation of the following matrix:

$$h \int_A \begin{bmatrix} N_i^2 & N_i N_j & N_i N_k \\ N_j N_i & N_j^2 & N_j N_k \\ N_k N_i & N_k N_j & N_k^2 \end{bmatrix} dA \qquad [4.27]$$

This matrix is used to include the effect of heat convection from the face area of the element, and appears in Eq. 6.85. Evaluate the matrix using natural coordinates.

Consider the first diagonal term, that is

$$\int N_i^2 \, dA = \int L_1^2 \, dA = \int L_1^2 L_2^0 L_3^0 \, dA = \frac{2! 0! 0!}{(2 + 0 + 0 + 2)!} 2A = \frac{A}{6}$$

(Note that factorial zero equals unity.) The other diagonal terms are similar, and give the same result. For the off-diagonal terms the integration involves, for example,

$$\int N_i N_j \, dA = \int L_1^1 L_2^1 L_3^0 \, dA = \frac{1! 1! 0!}{(1 + 1 + 0 + 2)!} 2A = \frac{A}{12}$$

Hence the integration in Eq. 4.27 may at first sight appear difficult, but with natural coordinates the integration becomes trivial. The final result is

$$\frac{hA}{12} \begin{bmatrix} 2 & 1 & 1 \\ 1 & 2 & 1 \\ 1 & 1 & 2 \end{bmatrix} \qquad [4.28]$$

Note that in Example 4.4 no information of the actual global coordinates of the element are used, so that the final Eq. 4.28 holds for all occurrences of that type of element. Therefore, for every triangular element that experiences face convection in a model, the above matrix must be added into the global stiffness matrix. The only information that is required is the area of the element, and of course the numbers of the nodes defining the element so that the matrix can be inserted into the global matrix in the correct position. This is a great strength of the finite element method. Once the format of the equations of a specific element type have been derived, it can be used for every occurrence of that element type.

Three-dimensional local coordinates

The application of natural coordinates to a three-dimensional simplex element is really of academic interest only, since the element is rarely used in practice. The basic shape introduced in Section 4.3 is a tetrahedron (Fig. 4.11). Not surprisingly, the natural coordinate for this type of element is a volume coordinate, and specifically the volume of a tetrahedron that forms part of the total element. Consider Fig. 4.14, which shows an element defined by nodes i, j, k and l, and an internal point P. The natural coordinates of the point are defined by

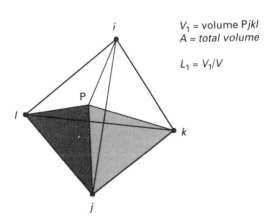

$V_1 = $ volume Pjkl
$A = $ total volume

$L_1 = V_1/V$

Figure 4.14 One of the natural (volume) coordinates of a three-dimensional element

$$L_1 = V_1/V \qquad L_2 = V_2/V \qquad L_3 = V_3/V \qquad L_4 = V_4/V \qquad [4.29]$$

where V is the total volume of the element, and V_1 for example is the volume of the tetrahedron Pjkl.

The integration formula that these natural coordinates then allow is

$$\int_V L_1^\alpha L_2^\beta L_3^\gamma L_4^\delta \, dV = \frac{\alpha! \beta! \gamma! \delta!}{(\alpha + \beta + \gamma + \delta + 3)!} 6V \qquad [4.30]$$

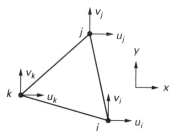

Figure 4.15 Degrees of freedom required to define a two-dimensional element for stress analysis problems

4.5   Vector quantities

So far, this chapter has only considered the variation of one quantity in each element, which is the case for field problems. The extension of the theories from a scalar to a vector quantity is however straightforward, with the components of the vector being considered individually. Therefore, for stress analysis problems, the displacements in the x, y and z directions are analysed independently, and each node then has up to three degrees of freedom, depending on the dimensionality of the problem. Consider the two-dimensional element shown in Fig. 4.15. It has a total of six degrees of freedom, where u and v are the displacements in the x and y directions respectively. The element requires two interpolation functions, one for each direction. In other words,

$$\begin{aligned} u &= N_i u_i + N_j u_j + N_k u_k \\ v &= N_i v_i + N_j v_j + N_k v_k \end{aligned} \qquad [4.31]$$

Since the shape functions are defined in Eq. 4.11 as

$$N_\beta = (a_\beta + b_\beta x + c_\beta y)/2A$$

where a, b and c are constants which depend solely on the geometry of the element (Eq. 4.8), the shape functions for the two interpolation functions must be the same.

The two equations of Eq. 4.28 can be stored most conveniently as

$$\begin{Bmatrix} u \\ v \end{Bmatrix} = \begin{bmatrix} N_i & 0 & N_j & 0 & N_k & 0 \\ 0 & N_i & 0 & N_j & 0 & N_k \end{bmatrix} \begin{Bmatrix} u_i \\ v_i \\ u_j \\ v_j \\ u_k \\ v_k \end{Bmatrix} = [N]\{U\} \qquad [4.32]$$

where $[N]$ is the shape function matrix and $\{U\}$ is a row vector of displacements.

For a three-dimensional element with four nodes, the shape function matrix will be of size $[3 \times 12]$, as given in the following:

$$\begin{Bmatrix} u \\ v \\ w \end{Bmatrix} = \begin{bmatrix} N_i & 0 & 0 & N_j & 0 & \dots & N_l & 0 & 0 \\ 0 & N_i & 0 & 0 & N_j & \dots & 0 & N_l & 0 \\ 0 & 0 & N_i & 0 & 0 & \dots & 0 & 0 & N_l \end{bmatrix} \begin{Bmatrix} u_i \\ v_i \\ w_i \\ u_j \\ \dots \\ \dots \\ w_l \end{Bmatrix} = [N]\{U\} \quad [4.33]$$

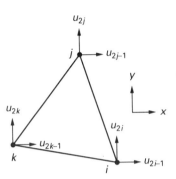

Figure 4.16 Alternative definitions for the degrees of freedom of a two-dimensional element

The computer does not of course differentiate between the u, v and w displacements – they are simply degrees of freedom – and therefore an alternative notation is often used. Figure 4.16 shows the system applied to the two-dimensional element. If the nodes of the element were 4, 6 and 10, then the degrees of freedom would be u_7, u_8, u_{11}, u_{12}, u_{19} and u_{20}. Hence the v displacement at node 10, formally known as v_{10}, is now labelled u_{20}. Using this system, the degrees of freedom are now labelled consecutively through the model. The degrees of freedom then fall naturally into the right place when a model is analysed and the system equations are developed. For the three-dimensional element, the degrees of freedom for node i would be defined as u_{3i-2}, u_{3i-1} and u_{3i}, and so on.

Example 4.5: alternative notation for displacements

For the mesh in Fig. 4.10, write down the interpolation functions of elements (3) and (4) for a stress analysis problem, using the simplified notation.

The nodes of element (3) are 6, 2 and 5; therefore the variation of u and v through the element would be given by

$$u^{(3)} = N_6^{(3)}u_{11} + N_2^{(3)}u_3 + N_5^{(3)}u_9$$
$$v^{(3)} = N_6^{(3)}u_{12} + N_2^{(3)}u_4 + N_5^{(3)}u_{10} \quad [4.34]$$

where the superscript (3) identifies the particular element. For element (4), the nodes are 3, 2 and 6; hence

$$u^{(4)} = N_3^{(4)}u_5 + N_2^{(4)}u_3 + N_6^{(4)}u_{11}$$
$$v^{(4)} = N_3^{(4)}u_6 + N_2^{(4)}u_4 + N_6^{(4)}u_{12} \quad [4.35]$$

Remember that the shape functions relating to the same node in the two elements are not equal; for example,

$$N_2^{(3)} \neq N_2^{(4)}$$

However, the displacements of the nodes (u_3, u_4 *etc.*) must be the same; hence they have no superscripts distinguishing them.

4.6   An axisymmetric element

The two-dimensional element first introduced in Section 4.3 can also be applied to axisymmetric problems. However, instead of operating in the x–y plane, the element is used in the r–z plane, where the z axis is the axis of symmetry (Fig. 4.17). When the element is rotated about the z axis it forms a torus as shown in the figure.

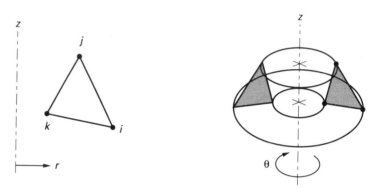

Figure 4.17 Axisymmetric element

Axisymmetric elements are defined as having a constant value of temperature or displacement (for example) in the circumferential or θ direction. This is analogous to two-dimensional problems where there is no variation in the out-of-plane direction, usually z. Therefore the interpolation function for the temperature distribution through the element can be written as

$$T = \alpha_1 + \alpha_2 r + \alpha_3 z \qquad\qquad [4.36]$$

If the α constants are evaluated in the usual way, the standard equation is derived which introduces the shape functions,

$$T = N_i T_i + N_j T_j + N_k T_k \qquad\qquad [4.37]$$

where the general form of the shape function is

$$N_\beta = (a_\beta + b_\beta r + c_\beta z)/2A \qquad \beta = i,j,k \qquad [4.38]$$

A comparison of this equation for the shape function shows that it is directly equivalent to the general Cartesian two-dimensional case in Eq. 4.11. The a, b and c constants are the same as those in Eq. 4.8, but with r and z coordinates replacing those in the x and y directions.

This element is used in Chapters 5 and 6 when the general equations for stress analysis and thermal problems are introduced.

4.7   Conclusions

This chapter introduces one of the fundamental concepts of the finite element method, the interpolation function, which describes how a field variable can be approximated across a finite element. Simplex elements with linear interpolation functions are discussed, and local coordinates are introduced to simplify the integration that is performed in the development of the system equations. More sophisticated elements are dealt with later in the book, and many of the concepts introduced here will be seen again.

One important property of interpolation functions has not been discussed in this chapter, and that is the convergence requirements of an element. These are the conditions ensuring that the element behaves properly and converges to the correct answer as its size decreases. Convergence requirements are investigated in Chapter 8, when all the steps of the finite element method have been introduced, and their implications can be more easily understood. For the time being it is reiterated that the method is approximate, but that the answer should converge to the correct solution as the number of elements is increased and consequently the size is decreased.

Some basic finite elements having been introduced, the next stage in the method is the development of the general governing equations of a particular problem type, into which the element details are substituted. This yields the element characteristic matrices and vectors, and ultimately the system equations for the whole problem. The derivation of the equations for stress analysis and field problems is discussed in the next two chapters.

Problems

4.1   Write down a cubic interpolation function for a triangular element. Sketch the element and suggest a suitable location for the 'extra' node.

4.2   Prove that the α constants for the two-dimensional simplex element are indeed given by Eq. 4.6.

4.3   Prove that the shape function N_i equals one at node i and zero at nodes j and k in a simplex triangular element.

4.4   Prove that the shape function N_i is zero along the side jk of a simplex triangular element. [Hint: write down the equation of the line defining the side jk, *i.e.* $y = mx + c$.]

4.5   Prove that the sum of the shape functions of the two-dimensional element defined by Eq. 4.10 is unity.

4.6   Calculate the shape functions for the elements in Fig. 4.18(a)–(d).

4.7   If the nodal temperatures of the element in Fig. 4.18(c) are found to be

$$\begin{Bmatrix} \phi_i \\ \phi_j \\ \phi_k \end{Bmatrix} = \begin{Bmatrix} 20 \\ 100 \\ 150 \end{Bmatrix} °C$$

what is the temperature at the point (2, 2)?

4.8   If the nodal displacements of the element in Fig. 4.18(d) are found to be

$$\begin{Bmatrix} u_i \\ v_i \\ u_j \\ v_j \\ u_k \\ v_k \end{Bmatrix} = \begin{Bmatrix} 2 \\ 2 \\ 1 \\ 0 \\ 3 \\ 1 \end{Bmatrix} \times 10^{-3} \text{ mm}$$

what are the displacements at the point (1, 3)?

4.9   Calculate the shape functions for element (4) in Fig 4.10, and show that $\phi^{(3)} = \phi^{(4)}$ along the common boundary of elements (3) and (4).

(a)

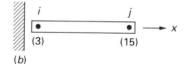

(b)

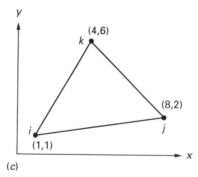

(c)

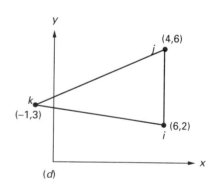

(d)

Figure 4.18

4.10   Integrate the following using local integration formulae:

(a) $\displaystyle\int_L N_i N_j \, dx$ (b) $\displaystyle\int_L N_i^2 \, dx$

(c) $\displaystyle\int_L N_i N_j^3 \, dx$

(d) $\displaystyle\int_A N_i N_j \, dA$ (e) $\displaystyle\int_A N_i N_j^2 N_k \, dA$

(f) $\displaystyle\int_V N_i N_j \, dV$ (g) $\displaystyle\int_V N_i^2 N_j \, dV$

(h) $\displaystyle\int_A \begin{Bmatrix} N_i \\ N_j \end{Bmatrix} [N_i \ N_j] \, dA$ (i) $\displaystyle\int_H \begin{Bmatrix} N_i \\ N_j \\ N_k \end{Bmatrix} [N_i \ N_j \ N_k] \, dH$

4.11   Prove the general rule that $\phi^{(1)} = \phi^{(2)}$ along the common boundary of the two triangular elements shown in Fig. 4.19. [Hint: use the fact that $N_i^{(e)} = L_1^{(e)}$ and $N_j^{(e)} = L_2^{(e)}$ for each element, and examine the values of the local coordinates along the boundary.]

4.12   Sketch a three-dimensional simplex element for stress analysis and mark on the degrees of freedom using the simplified notation scheme of Fig. 4.16.

4.13   If the nodes of the element defined in Problem 4.12 are 7, 3, 10 and 12, write down the interpolation functions for the three displacements in terms of the shape functions and nodal displacements.

4.14   Show that the condition to be satisfied for convergence of a field variable to a constant value in an element is that the sum of the shape functions equals unity at every point in the element.

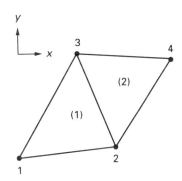

Figure 4.19

5 Formulation of the element characteristic matrices and vectors for elasticity problems

5.1 Introduction

The finite element method is used most frequently to analyse stress problems. The technique can readily cope with several types of applied loading, namely concentrated forces, distributed or pressure loads, body force loads (inertia or gravitational), initial strains (usually thermally induced) and prescribed displacements. As with all elasticity analyses, the principle of superposition applies, and a finite model of a body under any combination of loading conditions can be solved to yield the nodal displacements. From the displacements, it is then possible to calculate the strains and ultimately the stresses.

This chapter is concerned with the derivation of the finite element equations for elasticity problems. There are several methods that can be used, but the most straightforward approach is the minimization of the potential energy of the system. The procedure is known more generally as a variational formulation. The other most common technique of deriving the equations is by a weighted residual approach, in particular that due to Galerkin. For elasticity problems this second method is rarely used and it is not discussed in this chapter, but with field problems both techniques are frequently applied and consequently they are both discussed in the following chapter. As one might expect, the same equations are derived whichever approach is used.

The object of the steps in the next section is to derive the general element equations for stress analysis problems. These basic equations are then tailored specifically to one-, two- or three-dimensional or axisymmetric analyses.

5.2 The variational formulation

This method has already been introduced in Section 2.2, when the potential energy of a simple one-dimensional bar was minimized. The potential energy Π equals the strain energy Λ in the body less the work done W by the external loads acting on the body:

$$\Pi = \Lambda - W \tag{5.1}$$

For a differential element of volume dV, the strain energy is found from

$$d\Lambda = \frac{1}{2}\{\varepsilon\}^{\mathrm{T}}\{\sigma\} - \frac{1}{2}\{\varepsilon_0\}^{\mathrm{T}}\{\sigma\} \tag{5.2}$$

where $\{\varepsilon\}$ is a vector of the total strains; $\{\varepsilon_0\}$ is a vector of initial strain, for example from a temperature differential (typically $\varepsilon_0 = \alpha\Delta T$, where α is the coefficient of thermal expansion and ΔT is the change in temperature); and $\{\sigma\}$ is a vector of the stress components. It is important to note that $\{\varepsilon\}$ is the total strain, and includes the effects of any initial strains.

The total strain energy for a finite volume is therefore calculated by integrating Eq. 5.2 through the volume. Hence

$$\Lambda = \int_V \frac{1}{2}(\{\varepsilon\}^{\mathrm{T}}\{\sigma\} - \{\varepsilon_0\}^{\mathrm{T}}\{\sigma\})dV \tag{5.3}$$

Since stress and strain are related by the elastic constants, this can be simplified. For example, for a two-dimensional analysis

$$\begin{aligned}\{\varepsilon\}^{\mathrm{T}} &= \{\varepsilon_x \ \varepsilon_y \ \gamma_{xy}\} \\ \{\sigma\}^{\mathrm{T}} &= \{\sigma_x \ \sigma_y \ \tau_{xy}\}\end{aligned} \tag{5.4}$$

and

$$\{\sigma\} = [D]\{\varepsilon\} - [D]\{\varepsilon_0\} \tag{5.5}$$

where

$$[D] = \frac{E}{(1+\nu)(1-2\nu)}\begin{bmatrix} 1-\nu & \nu & 0 \\ \nu & 1-\nu & 0 \\ 0 & 0 & \frac{1-2\nu}{2} \end{bmatrix} \tag{5.6}$$

if the problem is a plane strain approximation.

Therefore, Eq. 5.5 may be substituted into Eq. 5.3 to give

$$\Lambda = \frac{1}{2}\int_V (\{\varepsilon\}^{\mathrm{T}}[D]\{\varepsilon\} - 2\{\varepsilon\}^{\mathrm{T}}[D]\{\varepsilon_0\} + \{\varepsilon_0\}^{\mathrm{T}}[D]\{\varepsilon_0\})\,dV \tag{5.7}$$

The finite element method divides the volume into a number of smaller regions or elements, and approximates the displacement throughout each element, by for example

$$\{u\} = [N]\{U\} \tag{5.8}$$

which was derived in Eqs 4.32 and 4.33.

The strains are simple functions of these displacements according to basic elasticity theory. For a full three-dimensional strain field,

$$\varepsilon_x = \frac{\partial u}{\partial x} \qquad \varepsilon_y = \frac{\partial v}{\partial y} \qquad \varepsilon_z = \frac{\partial w}{\partial z}$$

$$\gamma_{xy} = \frac{\partial u}{\partial y} + \frac{\partial v}{\partial x} \qquad \gamma_{yz} = \frac{\partial v}{\partial z} + \frac{\partial w}{\partial y} \qquad \gamma_{zx} = \frac{\partial w}{\partial x} + \frac{\partial u}{\partial z} \qquad [5.9]$$

If these strains are stored in a vector such as that in Eq. 5.4, then they can be equated to the nodal displacements by

$$\{\varepsilon\} = [B]\{U\} \qquad [5.10]$$

To determine the $[B]$ matrix, consider for example the displacement in the x direction in a two-dimensional element, which is given by Eq. 4.31 as

$$u = N_i u_i + N_j u_j + N_k u_k$$

where u_i, u_j and u_k are the nodal displacements in that direction and form part of $\{U\}$. The strain in the x direction is calculated from

$$\varepsilon_x = \frac{\partial u}{\partial x} = \frac{\partial N_i}{\partial x} u_i + \frac{\partial N_j}{\partial x} u_j + \frac{\partial N_k}{\partial x} u_k \qquad [5.11]$$

Hence $[B]$ is a matrix of terms derived by the proper differentiation of the shape functions. Consideration of the general format of the shape functions of two-dimensional simplex elements (Eq. 4.11) shows that this differentiation is trivial, resulting in a constant. For higher-order elements the calculation is similarly straightforward, but yields linear, quadratic or higher-order functions.

Substitution of Eq. 5.10 into Eq. 5.7 now gives the strain energy $\Lambda^{(e)}$ for a single element (e) as

$$\Lambda^{(e)} = \frac{1}{2} \int_{V^{(e)}} (\{U^{(e)}\}^{\mathrm{T}} [B^{(e)}]^{\mathrm{T}} [D^{(e)}][B^{(e)}]\{U^{(e)}\} - 2\{U^{(e)}\}^{\mathrm{T}} [B^{(e)}]^{\mathrm{T}} \\ [D^{(e)}]\{\varepsilon_0^{(e)}\} + \{\varepsilon_0^{(e)}\}^{\mathrm{T}} [D^{(e)}]\{\varepsilon_0^{(e)}\}) \, \mathrm{d}V \qquad [5.12]$$

The strain energy of the whole volume can then be found by summing the strain energies of all the individual elements.

The work done on an element by the applied loads can be separated into three parts, that is the work done by the distributed (pressure) loads, by the body (inertia) forces and by the concentrated nodal loads (Fig. 5.1). The work done by any force is the magnitude of the force multiplied by the distance moved. Therefore if the body forces are $\mathbb{X}$, $\mathbb{Y}$, $\mathbb{Z}$ per unit volume then, for a general element, the work done by the body forces is

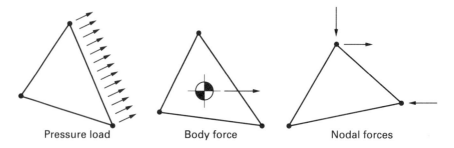

Figure 5.1 Types of applied
loads that a two-dimensional
element can experience

Pressure load   Body force   Nodal forces

$$W_B^{(e)} = \int_{V^{(e)}} (u\mathbb{X}^{(e)} + v\mathbb{Y}^{(e)} + w\mathbb{Z}^{(e)})\mathrm{d}V \qquad [5.13]$$

Using Eq. 5.8, this can be simplified to

$$W_B^{(e)} = \int_{V^{(e)}} \{U^{(e)}\}^{\mathrm{T}}[N^{(e)}]^{\mathrm{T}}\begin{Bmatrix} \mathbb{X}^{(e)} \\ \mathbb{Y}^{(e)} \\ \mathbb{Z}^{(e)} \end{Bmatrix} \mathrm{d}V \qquad [5.14]$$

The pressure forces will act over a surface, and consequently the work
done will be found from

$$W_P^{(e)} = \int_{S^{(e)}} (up_x^{(e)} + vp_y^{(e)} + wp_z^{(e)})\mathrm{d}S =$$
$$\int_{S^{(e)}} \{U^{(e)}\}^{\mathrm{T}}[N^{(e)}]^{\mathrm{T}} \begin{Bmatrix} p_x^{(e)} \\ p_y^{(e)} \\ p_z^{(e)} \end{Bmatrix} \mathrm{d}S \qquad [5.15]$$

where p_x, p_y and p_z are the distributed pressure loads parallel to the
coordinate axes (Fig. 5.2), and are considered positive when acting in the
positive coordinate direction.

The nodal loads are the most simple to deal with. If $\{P^{(e)}\}$ is a vector of
the nodal loads, then the work done will be

$$W_C^{(e)} = \{U^{(e)}\}^{\mathrm{T}}\{P^{(e)}\} \qquad [5.16]$$

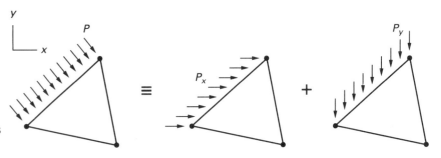

Figure 5.2 Resolution of an
applied pressure loading into its
Cartesian components

Therefore the potential energy of a single element is given by Eqs 5.12 and 5.14 to 5.16, namely

$$\Pi^{(e)} = \Lambda^{(e)} - W_{\mathrm{B}}^{(e)} - W_{\mathrm{P}}^{(c)} - W_{\mathrm{C}}^{(e)} \qquad [5.17]$$

The potential energy of the whole system is then the sum of the energies of all the elements in the model. More specifically,

$$
\begin{aligned}
\Pi = \sum_{e=1}^{E} \Bigg[&\int_{V^{(e)}} \frac{1}{2} \{U^{(e)}\}^{\mathrm{T}} [B^{(e)}]^{\mathrm{T}} [D^{(e)}][B^{(e)}]\{U^{(e)}\} \mathrm{d}V \\
&- \int_{V^{(e)}} \{U^{(e)}\}^{\mathrm{T}} [B^{(e)}]^{\mathrm{T}} [D^{(e)}]\{\varepsilon_0^{(e)}\} \mathrm{d}V \\
&+ \int_{V^{(e)}} \frac{1}{2} \{\varepsilon_0^{(e)}\}^{\mathrm{T}} [D^{(e)}]\{\varepsilon_0^{(e)}\} \mathrm{d}V \\
&- \int_{V^{(e)}} \{U^{(e)}\}^{\mathrm{T}} [N^{(e)}]^{\mathrm{T}} \left\{ \begin{array}{c} \mathbb{X}^{(e)} \\ \mathbb{Y}^{(e)} \\ \mathbb{Z}^{(e)} \end{array} \right\} \mathrm{d}V \\
&- \int_{S^{(e)}} \{U^{(e)}\}^{\mathrm{T}} [N^{(e)}]^{\mathrm{T}} \left\{ \begin{array}{c} p_x^{(e)} \\ p_y^{(e)} \\ p_z^{(e)} \end{array} \right\} \mathrm{d}S - \{U^{(e)}\}^{\mathrm{T}}\{P^{(e)}\} \Bigg] \qquad [5.18]
\end{aligned}
$$

Equation 5.18 is the total potential energy of a body undergoing various loading conditions. It is not as complex as it appears; if it were to be evaluated it would give a single number – the potential energy.

For the body to be in equilibrium, its potential energy must be a minimum. Hence if the y displacement of node 3, for example, is u_6, then

$$\frac{\partial \Pi}{\partial u_6} = 0$$

for the minimum to be obtained. The same applies to all the other degrees of freedom in the model. If there are n degrees of freedom,

$$\frac{\partial \Pi}{\partial u_1} = \frac{\partial \Pi}{\partial u_2} = \dots = \frac{\partial \Pi}{\partial u_n} = 0 \qquad [5.19]$$

This gives n equations in n unknowns (the degrees of freedom), and is expressed more conveniently as

$$\frac{\partial \Pi}{\partial \{U_{\mathrm{S}}\}} = 0 \qquad [5.20]$$

where $\{U_{\mathrm{S}}\}$ is the vector of all the degrees of freedom in the model, and consists of all the elemental displacement vectors. Therefore differentiation of Eq. 5.18 by the standard methods gives

$$\frac{\partial \Pi}{\partial \{U_S\}} = \sum_{e=1}^{E} \left[\int_{V^{(e)}} [B^{(e)}]^T [D^{(e)}][B^{(e)}] dV \right] \{U_S\}$$

$$- \sum_{e=1}^{E} \left[\int_{V^{(e)}} [B^{(e)}]^T [D^{(e)}]\{\varepsilon_0^{(e)}\} dV + \int_{V^{(e)}} [N^{(e)}]^T \left\{ \begin{array}{c} \mathbb{X}^{(e)} \\ \mathbb{Y}^{(e)} \\ \mathbb{Z}^{(e)} \end{array} \right\} dV \right.$$

$$\left. + \int_{S^{(e)}} [N^{(e)}]^T \left\{ \begin{array}{c} p_x^{(e)} \\ p_y^{(e)} \\ p_z^{(e)} \end{array} \right\} dS \right] - \{P_S\} = 0 \qquad [5.21]$$

where $\{P^{(e)}\}$ has been taken outside the summations and replaced by $\{P_S\}$, the complete vector of nodal loads.

Equation 5.21 may be written as

$$[k]\{U_S\} = \{F\}$$

where

$$[k] = \sum_{e=1}^{E} \left[\int_{V^{(e)}} [B^{(e)}]^T [D^{(e)}][B^{(e)}] dV \right] = \sum_{e=1}^{E} [k^{(e)}] \qquad [5.22]$$

and is known as the global (or system) matrix, and the global force vector is

$$\{F\} = \sum_{e=1}^{E} \left[\int_{V^{(e)}} [B^{(e)}]^T [D^{(e)}]\{\varepsilon_0^{(e)}\} dV + \int_{V^{(e)}} [N^{(e)}]^T \left\{ \begin{array}{c} \mathbb{X}^{(e)} \\ \mathbb{Y}^{(e)} \\ \mathbb{Z}^{(e)} \end{array} \right\} dV \right.$$

$$\left. + \int_{S^{(e)}} [N^{(e)}]^T \left\{ \begin{array}{c} p_x^{(e)} \\ p_y^{(e)} \\ p_z^{(e)} \end{array} \right\} dS \right] + \{P_S\} = \sum_{e=1}^{E} \{F^{(e)}\} \qquad [5.23]$$

If a model has n degrees of freedom, then $[k]$ will be an $[n \times n]$ square matrix, while a single element with m degrees of freedom will result in an $[m \times m]$ square matrix for $[k^{(e)}]$. The global force vector, on the other hand, will be a vector of length n, while the element vector will be of length m.

The finite element method consists of calculating the individual element stiffness matrices and vectors, and summing (or assembling) them into the global stiffness matrix and force vector. The set of simultaneous equations that this produces is then solved for the nodal displacements.

Equations 5.22 and 5.23 have the general form of the system equations for elasticity problems. These equations must now be tailored to specific problem types, for example two- or three-dimensional analyses, and these are discussed individually in the following sections. For simplicity the superscript (e) is dropped where only one element is being considered.

5.3   One-dimensional elasticity

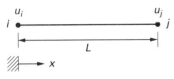

Figure 5.3 One-dimensional simplex element

A general one-dimensional (simplex) element is shown in Fig. 5.3. It models the x displacement, and consequently can only predict the σ_x stress. The displacement function for the element is

$$u = N_i u_i + N_j u_j = [N]\{U\} \qquad [5.24]$$

where the shape functions are

$$N_i = 1 - \frac{x}{L} \quad \text{and} \quad N_j = \frac{x}{L} \qquad [5.25]$$

The axial strain is found by differentiating Eq. 5.24 with respect to x:

$$\varepsilon_x = \frac{du}{dx} = \frac{dN_i}{dx} u_i + \frac{dN_j}{dx} u_j = \frac{1}{L}[-1\ 1]\begin{Bmatrix} u_i \\ u_j \end{Bmatrix}$$

Clearly then the $[B]$ matrix as defined by Eq. 5.10 must be

$$[B] = \frac{1}{L}[-1\ 1] \qquad [5.26]$$

For this simple element the stress and strain are related by

$$\sigma_x = E(\varepsilon_x - \varepsilon_0)$$

Hence $[D]$ equals E, Young's modulus, and the initial strain is usually $\alpha\Delta T$, a thermal strain.

Substituting $[B]$ and $[D]$ into Eq. 5.22 gives the stiffness matrix for the element. Assuming a constant cross-sectional area A,

$$[k] = \int_V [B]^T[D][B]dV$$

$$= \int_V \frac{1}{L}\begin{Bmatrix} -1 \\ 1 \end{Bmatrix} E \frac{1}{L}[-1\ 1]dV$$

$$= \frac{E}{L^2}\begin{bmatrix} 1 & -1 \\ -1 & 1 \end{bmatrix}\int_L A\ dx = \frac{AE}{L}\begin{bmatrix} 1 & -1 \\ -1 & 1 \end{bmatrix} \qquad [5.27]$$

This should be familiar to most engineers.

The integrals in the column force vector of Eq. 5.23 are also readily evaluated using the basic element information. For the thermal strain term,

$$\int_V [B]^T[D]\{\varepsilon_0\}dV = \int_V \frac{1}{L}[-1\ 1]^T E\alpha\Delta T\ dV$$

$$= \frac{E\alpha\Delta T}{L}\begin{Bmatrix} -1 \\ 1 \end{Bmatrix}\int_L A\ dx = EA\alpha\Delta T\begin{Bmatrix} -1 \\ 1 \end{Bmatrix} \qquad [5.28]$$

The body force term only considers an effect in the x direction, since all the other displacements are zero. Hence

$$\int_V [N]^T\{\mathbb{X}\} dV = \int_V \left\{ \begin{array}{c} N_i \\ N_j \end{array} \right\} \mathbb{X}\, dV$$

$$= \int_L \left\{ \begin{array}{c} N_i \\ N_j \end{array} \right\} \mathbb{X}A\, dx = \frac{\mathbb{X}AL}{2} \left\{ \begin{array}{c} 1 \\ 1 \end{array} \right\} \qquad [5.29]$$

Thus the body force is divided evenly between the two nodes.

The integrals of the shape functions are evaluated using the factorial integration formula of Eq. 4.20. Since $N_i = L_1$ and $N_j = L_2$, each line involves

$$\int_L N_i\, dx = \int_L L_1\, dx = \frac{1!}{2!}L = \frac{L}{2}$$

If a pressure of p_x is considered acting at node i of the element, then the term in Eq. 5.23 relating to the pressure will be

$$\int_S [N]^T\{p_x\} dS = \int_A \left\{ \begin{array}{c} N_i \\ N_j \end{array} \right\} p_x\, dA$$

However, at node i, $N_i = 1$ and $N_j = 0$; hence the integral becomes

$$\int_A \left\{ \begin{array}{c} 1 \\ 0 \end{array} \right\} p_x\, dA = p_x A \left\{ \begin{array}{c} 1 \\ 0 \end{array} \right\} \qquad [5.30]$$

A similar term would be derived if the pressure had been applied to node j with the 0 and 1 reversed.

Therefore the set of equations for a one-dimensional element with a pressure load at end i is

$$\underbrace{\frac{AE}{L} \left[\begin{array}{cc} 1 & -1 \\ -1 & 1 \end{array} \right] \left\{ \begin{array}{c} u_i \\ u_j \end{array} \right\}}_{\substack{\text{stiffness} \\ \text{matrix}}} = \underbrace{EA\alpha\Delta T \left\{ \begin{array}{c} -1 \\ 1 \end{array} \right\}}_{\substack{\text{from thermal} \\ \text{expansion}}} + \underbrace{Ap_x \left\{ \begin{array}{c} 1 \\ 0 \end{array} \right\}}_{\substack{\text{from pressure} \\ \text{on face } i}}$$

$$+ \underbrace{\frac{\mathbb{X}AL}{2} \left\{ \begin{array}{c} 1 \\ 1 \end{array} \right\}}_{\substack{\text{from body} \\ \text{forces}}} + \underbrace{\left\{ \begin{array}{c} P_i \\ P_j \end{array} \right\}}_{\substack{\text{nodal} \\ \text{forces}}} \qquad [5.31]$$

where P_i and P_j are the concentrated nodal forces. Note that this equation is derived by assuming that the element has a constant cross-sectional area, thus simplifying the calculations. If this is not the case, the variation of area with coordinate x needs to be determined and used in the integration.

Example 5.1: a bar with varying cross-sectional area

Calculate the stiffness matrix and body force vector for a finite element with a cross-sectional area that varies linearly down its length, as shown in Fig. 5.4.

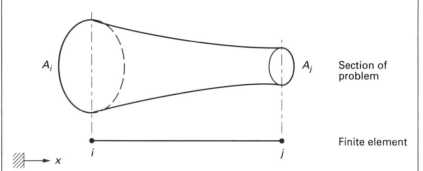

Section of problem

Finite element

Figure 5.4

The area is a linear function of x; hence

$$A = a + bx \tag{5.32}$$

If the areas are A_i and A_j at nodes i and j respectively, then

$$A_i = a + bx_i$$
$$A_j = a + bx_j$$

These two equations are solved for a and b, which are then substituted back into Eq. 5.32 to give

$$A = \frac{A_i x_j - A_j x_i}{L} + \frac{A_j - A_i}{L} x \tag{5.33}$$

Looking back to Eq. 4.2, it is seen that this is precisely the same form as the variation of the linear interpolation function. If Eq. 5.33 is reordered, it yields

$$A = \frac{x - x_i}{L} A_i + \frac{x_j - x}{L} A_j$$

or

$$A = N_i A_i + N_j A_j = [N] \begin{Bmatrix} A_i \\ A_j \end{Bmatrix} \tag{5.34}$$

where N_i and N_j are the standard one-dimensional shape functions. Therefore, since the displacements and cross-sectional areas of this tapered element are both linear functions, they use the same shape functions (Eqs 5.24 and 5.34).

To evaluate the governing equations of the element, Eq. 5.34 must be used rather than a constant value A. For the stiffness matrix, the derivation of Eq. 5.27 involved

$$[k] = \frac{E}{L^2}\begin{bmatrix} 1 & -1 \\ -1 & 1 \end{bmatrix}\int_L A\, dx$$

$$= \frac{E}{L^2}\begin{bmatrix} 1 & -1 \\ -1 & 1 \end{bmatrix}\int_L [N_i\ N_j]\begin{Bmatrix} A_i \\ A_j \end{Bmatrix} dx \qquad [5.35]$$

This can be evaluated using natural coordinates:

$$\int_L N_i\, dx = \int_L L_1\, dx = \frac{L}{2}$$

Therefore Eq. 5.35 becomes

$$[k] = \frac{E}{L^2}\begin{bmatrix} 1 & -1 \\ -1 & 1 \end{bmatrix}\begin{bmatrix} \dfrac{L}{2} & \dfrac{L}{2} \end{bmatrix}\begin{Bmatrix} A_i \\ A_j \end{Bmatrix}$$

$$= \frac{(A_i + A_j)E}{2L}\begin{bmatrix} 1 & -1 \\ -1 & 1 \end{bmatrix} \qquad [5.36]$$

This is similar to Eq. 5.27 but with the average cross-sectional area replacing the constant value.

For the body force term, the calculation of Eq. 5.29 involves

$$\int_V [N]^T\{\mathbb{X}\}dV = \int_L \begin{Bmatrix} N_i \\ N_j \end{Bmatrix}\mathbb{X}[N_i\ N_j]\begin{Bmatrix} A_i \\ A_j \end{Bmatrix} dx$$

$$= \int_L \begin{bmatrix} N_i^2 & N_iN_j \\ N_jN_i & N_j^2 \end{bmatrix}\begin{Bmatrix} A_i \\ A_j \end{Bmatrix}\mathbb{X}\, dx = \frac{\mathbb{X}L}{6}\begin{bmatrix} 2 & 1 \\ 1 & 2 \end{bmatrix}\begin{Bmatrix} A_i \\ A_j \end{Bmatrix}$$

$$[5.37]$$

Clearly this is not simply a case of using the average cross-sectional area. The end with the larger cross-section is given a larger share of the force.

The pin-jointed bar in two- and three-dimensional space

At the beginning of Section 5.3 and in Section 2.2, the stiffness matrix for a horizontal pin-jointed bar element has been developed and found to be (Eq. 5.27)

$$[k] = \frac{AE}{L}\begin{bmatrix} 1 & -1 \\ -1 & 1 \end{bmatrix}$$

However, in general this is not very useful since most members in practical trusses will be inclined to the horizontal. So the stiffness matrix

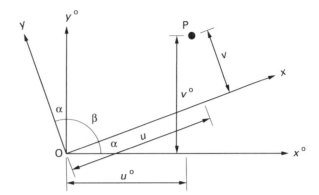

Figure 5.5 Relationship between local (element) and global coordinate systems

needs to be converted to two- and three-dimensional forms, which are developed below. The procedure is not difficult once the relationship between the local (element) and global coordinates has been derived.

Consider the general case in Fig. 5.5, where a point moves from the origin O to location P. The local axes are defined by x and y, with displacements of u and v, while the global terms have superscripts, *i.e.* x^o, y^o, u^o and v^o. The local and global displacements are related by

$$\begin{Bmatrix} u \\ v \end{Bmatrix} = \begin{bmatrix} \cos \alpha \cos \beta \\ -\cos \beta \cos \alpha \end{bmatrix} \begin{Bmatrix} u^o \\ v^o \end{Bmatrix} = \begin{bmatrix} l & m \\ -m & l \end{bmatrix} \begin{Bmatrix} u^o \\ v^o \end{Bmatrix} \qquad [5.38]$$

where $l = \cos \alpha$ and $m = \cos \beta$ are the direction cosines of the local x axis (along which the one-dimensional bar element is assumed to lie). This may be written as

$$\{U_i\} = [\xi]\{U_i^o\}$$

if the displacements are those of node i in a truss element.

Since the element has two nodes, the total displacement vector is given by

$$\begin{Bmatrix} u_i \\ v_i \\ u_j \\ v_j \end{Bmatrix} = \begin{bmatrix} \xi & 0 \\ \hline 0 & \xi \end{bmatrix} \begin{Bmatrix} u_i^o \\ v_i^o \\ u_j^o \\ v_j^o \end{Bmatrix} \qquad [5.39]$$

or

$$\{U\} = [\lambda]\{U^o\} \qquad [5.40]$$

The same type of coordinate transformation will apply to the terms of the element force vector, so that

$$\{F\} = [\lambda]\{F^o\} \qquad [5.41]$$

where

$$\{F\} = \{F_{ix} \ F_{iy} \ F_{jx} \ F_{jy}\}$$
$$\{F^o\} = \{F^o_{ix} \ F^o_{iy} \ F^o_{jx} \ F^o_{jy}\}$$

[5.42]

with F_{ix} and F_{iy} being the forces at node i in the x and y directions in the local coordinate system, and so on.

Since

$$\{F\} = [k]\{U\}$$

[5.43]

substitution of Eqs 5.40 and 5.41 gives

$$[\lambda]\{F^o\} = [k][\lambda]\{U^o\}$$

[5.44]

However, $[\lambda]$ is orthogonal, so that $[\lambda]^T = [\lambda]^{-1}$ and Eq. 5.44 can be rewritten as

$$\{F^o\} = [\lambda]^T[k][\lambda]\{U^o\}$$

[5.45]

Consequently, the elemental stiffness matrix in two-dimensional global coordinates is given by

$$[k^o] = [\lambda]^T[k][\lambda]$$

[5.46]

The stiffness matrix in Eq. 5.27 for the one-dimensional element only considers displacements in the x direction, since the forces and displacements in the y direction are assumed to be zero. Equation 5.39 can therefore be simplified to

$$\left\{ \begin{array}{c} u_i \\ u_j \end{array} \right\} = \left[\begin{array}{cccc} l & m & 0 & 0 \\ 0 & 0 & l & m \end{array} \right] \left\{ \begin{array}{c} u^o_i \\ v^o_i \\ u^o_j \\ v^o_j \end{array} \right\} = [\lambda]\{U^o\}$$

[5.47]

Now if Eq. 5.46 is expanded using Eqs 5.27 and 5.47, it is finally shown that

$$[k^o] = \frac{AE}{L} \left[\begin{array}{cccc} l^2 & lm & - l^2 & - lm \\ lm & m^2 & - lm & - m^2 \\ - l^2 & - lm & l^2 & lm \\ - lm & - m^2 & lm & m^2 \end{array} \right]$$

[5.48]

Note that if a bar is defined by nodes i and j, with global coordinates (x^o_i, y^o_i) and (x^o_j, y^o_j), then

$$L = \sqrt{[(x^o_j - x^o_i)^2 + (y^o_j - y^o_i)^2]}$$
$$l = (x^o_j - x^o_i)/L$$
$$m = (y^o_j - y^o_i)/L$$

[5.49]

where node i defines the origin of the local coordinate system, and the x axis is then defined by the line connecting it to node j.

Furthermore, any truss analysis using this simple two-dimensional element will result, in the first instance, in the derivation of the nodal displacements in global coordinates. If the axial forces in the bars are needed, Eq. 5.47 must be used to calculate the displacements in the local coordinates of the bar, and then Eq. 5.43 used to find the element force vector, or by applying Hooke's law to the axial displacements.

Example 5.2: simple two-dimensional truss

Find the axial force in the two members of the simple truss shown in Fig. 5.6, using the finite element method (working in mm).

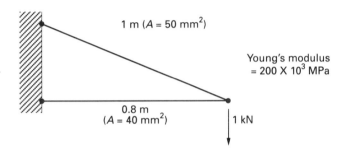

Figure 5.6

The finite element model is simply as shown in Fig. 5.7, and the element details are as follows:

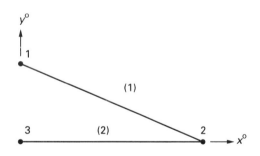

Figure 5.7

Element	i node	Coordinates x	y	j node	Coordinates x	y	L	l	m
(1)	1	0	600	2	800	0	1000	0.8	−0.6
(2)	3	0	0	2	800	0	800	1.0	0.0

where l and m are defined in Eq. 5.49.

Therefore the elemental stiffness matrices in global coordinates are found from Eq. 5.48 to be

$$[k^{o(1)}] = \frac{50 \times 2 \times 10^5}{1000} \begin{bmatrix} 0.64 & -0.48 & -0.64 & 0.48 \\ -0.48 & 0.36 & 0.48 & -0.36 \\ -0.64 & 0.48 & 0.64 & -0.48 \\ 0.48 & -0.36 & -0.48 & 0.36 \end{bmatrix}$$

$$[k^{o(2)}] = \frac{40 \times 2 \times 10^5}{800} \begin{bmatrix} 1 & 0 & -1 & 0 \\ 0 & 0 & 0 & 0 \\ -1 & 0 & 1 & 0 \\ 0 & 0 & 0 & 0 \end{bmatrix}$$

and the global stiffness matrix is

$$[k^o] = \begin{bmatrix} 0.64 & -0.48 & -0.64 & 0.48 & 0 & 0 \\ -0.48 & 0.36 & 0.48 & -0.36 & 0 & 0 \\ -0.64 & 0.48 & 1.64 & -0.48 & -1 & 0 \\ 0.48 & -0.36 & -0.48 & 0.36 & 0 & 0 \\ 0 & 0 & -1 & 0 & 1 & 0 \\ 0 & 0 & 0 & 0 & 0 & 0 \end{bmatrix} \times 10^4$$

The boundary conditions for the problem state that the displacements at nodes 1 and 3 are zero, and -1000 N is applied in the y direction at node 2. So

$$10^4 \times \begin{bmatrix} 0.64 & -0.48 & -0.64 & 0.48 & 0 & 0 \\ -0.48 & 0.36 & 0.48 & -0.36 & 0 & 0 \\ -0.64 & 0.48 & 1.64 & -0.48 & -1 & 0 \\ 0.48 & -0.36 & -0.48 & 0.36 & 0 & 0 \\ 0 & 0 & -1 & 0 & 1 & 0 \\ 0 & 0 & 0 & 0 & 0 & 0 \end{bmatrix} \begin{Bmatrix} 0 \\ 0 \\ u_2^o \\ v_2^o \\ 0 \\ 0 \end{Bmatrix}$$

$$= \begin{Bmatrix} 0 + R_{1x}^o \\ 0 + R_{1y}^o \\ 0 \\ -1000 \\ 0 + R_{3x}^o \\ 0 + R_{3y}^o \end{Bmatrix}$$

where the Rs are the reactions at the constrained nodes that must be applied to maintain the zero displacement.

Solving the third and fourth equations for the unknown displacements gives

$$u_2^o = -0.133 \text{ mm}$$
$$v_2^o = -0.456 \text{ mm}$$

To convert these into local coordinates for element (1), Eq. 5.38 is used, so that

$$\left\{ \begin{array}{c} u_2^{(1)} \\ v_2^{(1)} \end{array} \right\} = \left[\begin{array}{cc} 0.8 & -0.6 \\ 0.6 & 0.8 \end{array} \right] \left\{ \begin{array}{c} -0.133 \\ -0.456 \end{array} \right\} = \left\{ \begin{array}{c} 0.167 \\ -0.455 \end{array} \right\} \text{mm}$$

The force in element (1) by Hooke's law must be

$$F^{(1)} = AE \, (u_2^{(1)} - u_1^{(1)})/L = 1.67 \times 10^3 \text{ N}$$

and since the local coordinates of element (2) coincide with the global coordinates, for element (2)

$$F^{(2)} = AE \, (u_2^{(2)} - u_3^{(2)})/L = -1.33 \times 10^3 \text{ N}$$

These values agree exactly with those calculated theoretically.

Three-dimensional space trusses

A three-dimensional bar is processed in a similar way to the previous case. This time, however, three direction cosines l, m and n define the orientation of the bar in space, and

$$\left\{ \begin{array}{c} u_i \\ u_j \end{array} \right\} = \left[\begin{array}{cccccc} l & m & n & 0 & 0 & 0 \\ 0 & 0 & 0 & l & m & n \end{array} \right] \left\{ \begin{array}{c} u_i^o \\ v_i^o \\ w_i^o \\ u_j^o \\ v_j^o \\ w_j^o \end{array} \right\}$$

[5.50]

defines the axial displacement of the bar (in local coordinates) in terms of the global displacements.

It turns out that when the global stiffness matrix is calculated, it takes the form

$$[k^o] = \left[\begin{array}{c|c} \eta & -\eta \\ \hline -\eta & \eta \end{array} \right]$$

[5.51]

where

$$[\eta] = \frac{AE}{L} \left[\begin{array}{ccc} l^2 & lm & ln \\ ml & m^2 & mn \\ nl & nm & n^2 \end{array} \right]$$

[5.52]

$$l = (x_j^o - x_i^o)/L$$
$$m = (y_j^o - y_i^o)/L$$
$$n = (z_j^o - z_i^o)/L$$

[5.53]

with

$$L = \sqrt{[(x_j^o - x_i^o)^2 + (y_j^o - y_i^o)^2 + (z_j^o - z_i^o)^2]}$$

5.4   Two-dimensional elasticity

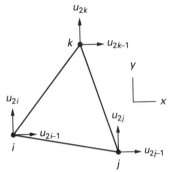

Figure 5.8 Nodal displacements of a two-dimensional simplex stress element

A two-dimensional problem will be either a plane stress or a plane strain approximation to a real three-dimensional situation, as discussed in Section 3.1. In either case, however, the same element is used. The difference occurs in the material property matrix $[D]$ and any initial strain conditions that might be experienced.

The two-dimensional simplex element is a triangle as shown in Fig. 5.8; it has two degrees of freedom at each node, which are labelled consecutively in the manner used by the computer. Following the procedures of the previous section, the $[B]$ and $[D]$ matrices are determined first for substitution into Eqs 5.22 and 5.23. The variations of the displacements in the x and y directions are given by Eq. 4.32 as

$$\begin{Bmatrix} u \\ v \end{Bmatrix} = \begin{bmatrix} N_i & 0 & N_j & 0 & N_k & 0 \\ 0 & N_i & 0 & N_j & 0 & N_k \end{bmatrix} \begin{Bmatrix} u_{2i-1} \\ u_{2i} \\ u_{2j-1} \\ u_{2j} \\ u_{2k-1} \\ u_{2k-1} \end{Bmatrix} = [N]\{U\}$$

where the general form of the shape functions is given by Eq. 4.11 as

$$N_\beta = \frac{1}{2A}(a_\beta + b_\beta x + c_\beta y) \qquad \beta = i,j,k$$

Remember that the a, b and c are constants and are calculated from the coordinates of the nodes.

The $[B]$ matrix is derived by the differentiation of the shape functions as discussed in Section 5.2. If the model is developed in the x–y plane, then

$$\gamma_{yz} = \gamma_{zx} = 0$$

Furthermore, for a plane stress approximation $\sigma_z = 0$, while $\varepsilon_z = 0$ when a plane strain approximation is used. Hence for both cases

$$\{\varepsilon\} = \begin{Bmatrix} \varepsilon_x \\ \varepsilon_y \\ \gamma_{xy} \end{Bmatrix} = [B]\{U\} \qquad\qquad [5.54]$$

where

$$\varepsilon_x = \frac{\partial u}{\partial x} \qquad \varepsilon_y = \frac{\partial v}{\partial y} \qquad \gamma_{xy} = \frac{\partial u}{\partial y} + \frac{\partial v}{\partial x}$$

Therefore, consider for example the strain in the x direction using Eqs 4.32 and 4.11:

$$\varepsilon_x = \frac{\partial u}{\partial x} = \left[\frac{\partial N_i}{\partial x} 0 \frac{\partial N_j}{\partial x} 0 \frac{\partial N_k}{\partial x} 0 \right] \{U\}$$

$$= \frac{1}{2A} [b_i \ 0 \ b_j \ 0 \ b_k \ 0] \{U\}$$

The other two strains can be found in the same way, and they can all be written into a matrix as

$$\begin{Bmatrix} \varepsilon_x \\ \varepsilon_y \\ \gamma_{xy} \end{Bmatrix} = \frac{1}{2A} \begin{bmatrix} b_i & 0 & b_j & 0 & b_k & 0 \\ 0 & c_i & 0 & c_j & 0 & c_k \\ c_i & b_i & c_j & b_j & c_k & b_k \end{bmatrix} \begin{Bmatrix} u_{2i-1} \\ u_{2i} \\ u_{2j-1} \\ u_{2j} \\ u_{2k-1} \\ u_{2k} \end{Bmatrix} = [B]\{U\} \quad [5.55]$$

Hence the $[B]$ matrix is of size $[6 \times 3]$ for a two-dimensional simplex element, and is composed only of constants.

If the element is to be a plane stress type, then

$$[D] = \frac{E}{1 + \nu^2} \begin{bmatrix} 1 & \nu & 0 \\ \nu & 1 & 0 \\ 0 & 0 & \dfrac{1 - \nu}{2} \end{bmatrix} \quad [5.56]$$

whereas a plane strain element will have

$$[D] = \frac{E}{(1 + \nu)(1 - 2\nu)} \begin{bmatrix} 1 - \nu & \nu & 0 \\ \nu & 1 - \nu & 0 \\ 0 & 0 & \dfrac{1 - 2\nu}{2} \end{bmatrix} \quad [5.57]$$

The other difference between plane stress and plane strain approximations occurs in the induced thermal strain vector. When the element is plane stress $\sigma_z = 0$, and consequently

$$\{\varepsilon_0\} = \alpha\Delta T \begin{Bmatrix} 1 \\ 1 \\ 0 \end{Bmatrix} \quad [5.58]$$

However, with a plane strain formulation $\varepsilon_z = 0$ but $\sigma_z \neq 0$, so that the initial strain vector becomes

$$\{\varepsilon_0\} = (1 + \nu)\alpha\Delta T \begin{Bmatrix} 1 \\ 1 \\ 0 \end{Bmatrix} \quad [5.59]$$

Now the element equations can be developed. Firstly, the stiffness matrix is calculated from

$$[k] = \int_V [B]^{\mathrm{T}}[D][B]\mathrm{d}V = [B]^{\mathrm{T}}[D][B]\int_V \mathrm{d}V$$

or

$$[k] = [B]^{\mathrm{T}}[D][B]tA \qquad [5.60]$$

The $[B]$ and $[D]$ matrices can be removed from the integration because they only contain constants, and the integral $\int \mathrm{d}V$ is replaced by tA, where t is the (constant) thickness of the element and A is the cross-sectional area. Equation 5.60 can then be expanded using Eqs 5.55 and either 5.56 or 5.57. The resulting matrix will be of size $[6 \times 6]$.

The force vector is defined in Eq. 5.23. Considering the thermal strain term first,

$$\int_V [B]^{\mathrm{T}}[D]\{\varepsilon_0\}\mathrm{d}V = [B]^{\mathrm{T}}[D]\{\varepsilon_0\}tA$$

For example, if the analysis is a plane stress approximation, the integral becomes

$$\frac{\alpha Et(\Delta T)}{2(1 - v)}\begin{Bmatrix} b_i \\ c_i \\ b_j \\ c_j \\ b_k \\ c_k \end{Bmatrix} \qquad [5.61]$$

The body force term is found from

$$\int_V [N]^{\mathrm{T}}\begin{Bmatrix} X \\ Y \end{Bmatrix}\mathrm{d}V = \int_A \begin{bmatrix} N_i & 0 \\ 0 & N_i \\ N_j & 0 \\ 0 & N_j \\ N_k & 0 \\ 0 & N_k \end{bmatrix}\begin{Bmatrix} X \\ Y \end{Bmatrix}t\,\mathrm{d}A = \frac{At}{3}\begin{Bmatrix} X \\ Y \\ X \\ Y \\ X \\ Y \end{Bmatrix} \qquad [5.62]$$

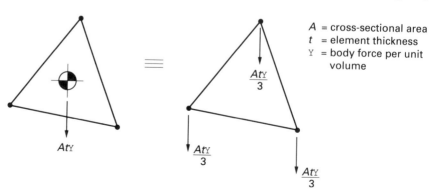

Figure 5.9 Distribution of a body force between the three nodes

A = cross-sectional area
t = element thickness
Y = body force per unit volume

Note that this term effectively distributes the body forces equally between the three nodes. For example, Fig. 5.9 shows how the effects of a vertical acceleration are dealt with.

Any pressures acting on the element must act in the plane of the element, and along one or more of the sides of the element, as previously shown in Fig. 5.2. Consider for example a pressure load acting over side ij of the element. The force vector term is

$$\int_{S_{ij}} [N]^T \begin{Bmatrix} p_x \\ p_y \end{Bmatrix} dS = \int_{S_{ij}} \begin{bmatrix} N_i & 0 \\ 0 & N_i \\ N_j & 0 \\ 0 & N_j \\ N_k & 0 \\ 0 & N_k \end{bmatrix} \begin{Bmatrix} p_x \\ p_y \end{Bmatrix} dS$$

where S is the surface over which the pressure acts. Since the element is of thickness t, this may be replaced by $H_{ij}t$, where H_{ij} is the length of the element's side between nodes i and j. Also the shape function matrix can be simplified, because along side ij the shape function $N_k = 0$. Making these substitutions and using the factorial integration formula of Eq. 4.25, the calculation proceeds in the following way:

$$\int_{S_{ij}} \begin{bmatrix} N_i & 0 \\ 0 & N_i \\ N_j & 0 \\ 0 & N_j \\ 0 & 0 \\ 0 & 0 \end{bmatrix} \begin{Bmatrix} p_x \\ p_y \end{Bmatrix} dS$$

$$= \int_{H_{ij}} \begin{bmatrix} N_i & 0 \\ 0 & N_i \\ N_j & 0 \\ 0 & N_j \\ 0 & 0 \\ 0 & 0 \end{bmatrix} \begin{Bmatrix} p_x \\ p_y \end{Bmatrix} t \, dH = \frac{H_{ij}t}{2} \begin{Bmatrix} p_x \\ p_y \\ p_x \\ p_y \\ 0 \\ 0 \end{Bmatrix} \qquad [5.63]$$

Thus the finite element method deals with pressure loads by applying equivalent nodal forces to the model. Clearly for this element, the total pressure load in each direction is divided equally between the two nodes on each face, as in Fig. 5.10.

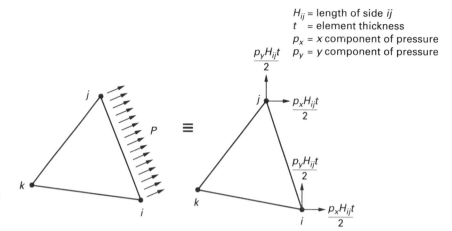

H_{ij} = length of side ij
t = element thickness
p_x = x component of pressure
p_y = y component of pressure

Figure 5.10 Example showing how pressures are included in the finite element models

This completes the derivation of the element equations for the two-dimensional element. Combining Eqs 5.61 to 5.63, and including a nodal force vector term, then for a plane stress element the total force vector is

$$\frac{\alpha Et(\Delta T)}{2(1-\nu)}\begin{Bmatrix} b_i \\ c_i \\ b_j \\ c_j \\ b_k \\ c_k \end{Bmatrix} + \frac{At}{3}\begin{Bmatrix} X \\ Y \\ X \\ Y \\ X \\ Y \end{Bmatrix} + \frac{H_{ij}t}{2}\begin{Bmatrix} p_x \\ p_y \\ p_x \\ p_y \\ 0 \\ 0 \end{Bmatrix} + \{P\} \qquad [5.64]$$

thermal expansion   body forces   pressure on side ij nodal forces

This includes just one term for a distributed load, on side ij. If pressures act on either of the other two faces, similar terms are added to the vector, with appropriate adjustment of the coefficients.

Example 5.3: two-dimensional plane stress analysis

Calculate the finite element equations of the two-dimensional plane stress element shown in Fig. 5.11.

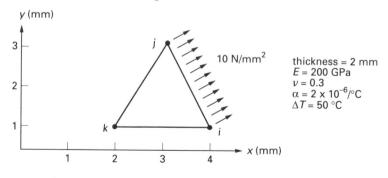

10 N/mm²

thickness = 2 mm
E = 200 GPa
ν = 0.3
α = 2 x 10⁻⁶/°C
ΔT = 50 °C

Figure 5.11

The [B] matrix for such a two-dimensional element is defined in Eq. 5.55. It comprises constants from the shape function equations, and these need to be calculated for each node. However, the element in Fig. 5.11 is identical to element (3) of Fig. 4.10, whose shape functions have been calculated in Example 4.1. Remembering that the shape functions depend solely on the geometry of the element, it can be concluded that

$$
\begin{aligned}
N_i &= 0.25(-3 + 2x - y) \\
N_j &= 0.25(-2 + 2y) \\
N_k &= 0.25(9 - 2x - y)
\end{aligned}
$$
[5.65]

These are taken from Eq. 4.12, with the nodes defined as in Fig. 5.11. The general form of the shape function is

$$
N_\beta = \frac{1}{2A}(a_\beta + b_\beta x + c_\beta y)
$$

Using Eq. 5.55,

$$
[B] = \frac{1}{4}
\begin{bmatrix}
2 & 0 & 0 & 0 & -2 & 0 \\
0 & -1 & 0 & 2 & 0 & -1 \\
-1 & 2 & 2 & 0 & -1 & -2
\end{bmatrix}
$$
[5.66]

The material property matrix is calculated according to Eq. 5.56:

$$
[D] = 0.22 \times 10^6
\begin{bmatrix}
1 & 0.3 & 0 \\
0.3 & 1 & 0 \\
0 & 0 & 0.35
\end{bmatrix}
$$
[5.67]

If Eqs 5.66 and 5.67 are used to calculate the stiffness matrix according to Eq. 5.60, then ultimately

$$
[k] =
\begin{bmatrix}
59.81 & -17.88 & -9.63 & 16.50 & -50.19 & 1.38 \\
-17.88 & 33.00 & 19.25 & -27.50 & -1.38 & -5.50 \\
-9.63 & 19.25 & 19.25 & 0.00 & -9.63 & -19.25 \\
16.50 & -27.50 & 0.00 & 55.00 & -16.50 & -27.50 \\
-50.19 & -1.38 & -9.63 & -16.50 & 59.81 & 17.88 \\
1.38 & -5.50 & -19.25 & -27.50 & 17.88 & 33.00
\end{bmatrix} \times 10^3
$$
[5.68]

Three simple checks can be performed after the calculations to detect any computational errors. Firstly, the matrix should be symmetric; secondly, the diagonal coefficients should all be non-zero and positive; and lastly, the sum of the terms in any one row or column should equal zero. The matrix above satisfies all these conditions.

The force vector for the given element will only contain terms from the pressure loading and the thermal strain. However, before Eq. 5.63 can be applied, the loading needs to be resolved into Cartesian components:

$$p_x = 10 \times \sin(63.43) = 8.94 \text{ N/mm}^2$$
$$p_y = 10 \times \cos(63.43) = 4.47 \text{ N/mm}^2$$

The length of the element side between nodes i and j is 2.24 mm, and therefore the force vector from Eq. 5.63 is

$$\frac{2.24 \times 2}{2} \left\{ \begin{array}{c} 8.94 \\ 4.47 \\ 8.94 \\ 4.47 \\ 0.0 \\ 0.0 \end{array} \right\} = \left\{ \begin{array}{c} 20.03 \\ 10.01 \\ 20.03 \\ 10.01 \\ 0.0 \\ 0.0 \end{array} \right\} \qquad [5.69]$$

For the thermal strain, Eq. 5.61 is used:

$$\frac{\alpha E t(\Delta T)}{2(1 - \nu)} \left\{ \begin{array}{c} b_i \\ c_i \\ b_j \\ c_j \\ b_k \\ c_k \end{array} \right\} = 28.57 \left\{ \begin{array}{c} 2 \\ -1 \\ 0 \\ 2 \\ -2 \\ -1 \end{array} \right\} \qquad [5.70]$$

The total force vector is then the sum of Eqs 5.69 and 5.70, which gives

$$\{F\} = \left\{ \begin{array}{c} 77.17 \\ -18.56 \\ 20.03 \\ 67.15 \\ -57.14 \\ -28.57 \end{array} \right\} \qquad [5.71]$$

Therefore the finite element equations for the specified element are given by Eqs 5.68 and 5.71.

Once a finite element model has been used to calculate the in-plane strains and stresses of two-dimensional problems, the out-of-plane strain ε_z or stress σ_z can be evaluated using standard elasticity equations if required.

For plane stress problems $\sigma_z = 0$, but the total out-of-plane strain is calculated from

$$\varepsilon_z = -\frac{v}{E}(\sigma_x + \sigma_y) + \alpha\Delta T \qquad [5.72]$$

When a plane strain problem is analysed $\varepsilon_z = 0$, but the out-of-plane stress is

$$\sigma_z = v(\sigma_x + \sigma_y) - E\alpha\Delta T \qquad [5.73]$$

Example 5.4: calculation of stresses from displacements

If the analysis of the element in Fig. 5.11 gives the following displacements, calculate the stresses predicted in the element, and the strain in the z direction:

Node	Displacements (10^{-3} mm)	
	u	v
i	2.0	0.0
j	0.5	2.0
k	1.0	−0.5

The total strains are calculated using Eq. 5.55:

$$\{\varepsilon\} = [B]\{U\}$$

From Eq. 5.66, this equals

$$
\left\{ \begin{array}{c} \varepsilon_x \\ \varepsilon_y \\ \varepsilon_{xy} \end{array} \right\} = \frac{1}{4}
\begin{bmatrix}
2 & 0 & 0 & 0 & -2 & 0 \\
0 & -1 & 0 & 2 & 0 & -1 \\
-1 & 2 & 2 & 0 & -1 & -2
\end{bmatrix}
\left\{ \begin{array}{c} 2.0 \\ 0.0 \\ 0.5 \\ 2.0 \\ 1.0 \\ -0.5 \end{array} \right\} \times 10^{-3}
$$

$$
= \left\{ \begin{array}{c} 500 \\ 1125 \\ -250 \end{array} \right\} \times 10^{-6}
$$

This is the total strain experienced by the element; before the stresses can be calculated, the values must be reduced by the thermal strains. According to Eq. 5.58, for a plane stress problem the thermal strains are

$$
\alpha\Delta T \left\{ \begin{array}{c} 1 \\ 1 \\ 0 \end{array} \right\} = \left\{ \begin{array}{c} 100 \\ 100 \\ 0 \end{array} \right\} \times 10^{-6}
$$

Therefore the strains due to the applied pressure loading are

$$\left\{ \begin{array}{c} 400 \\ 1025 \\ -250 \end{array} \right\} \times 10^{-6}$$

and the induced stresses are calculated using Eq. 5.56 as

$$\left\{ \begin{array}{c} \sigma_x \\ \sigma_y \\ \sigma_{xy} \end{array} \right\} = \frac{E}{1-v^2} \left[\begin{array}{ccc} 1 & v & 0 \\ v & 1 & 0 \\ 0 & 0 & \frac{1-v}{2} \end{array} \right] \left\{ \begin{array}{c} 400 \\ 1025 \\ -250 \end{array} \right\} \times 10^{-6}$$

$$= \left\{ \begin{array}{c} 156 \\ 252 \\ -19 \end{array} \right\} \text{MPa}$$

Finally, the total strain in the z direction is given by

$$\varepsilon_z = -\frac{v}{E}(\sigma_x + \sigma_y) + \alpha \Delta T$$

$$= -\frac{0.3}{200 \times 10^3}(156 + 252) + 100 \times 10^{-6} = -512 \times 10^{-6}$$

Note that the calculations yield a constant value for each stress and strain in the element. This is a major drawback of these simplex stress elements, and arises because the interpolation function of the displacements is linear. Clearly then, for areas with a rapidly changing stress field, a fine mesh is required to reasonably approximate the stress variation. Consider for example Fig. 5.12, which shows how a stress field could be modelled by a number of different sized simplex elements, resulting in a gradation of the mesh density.

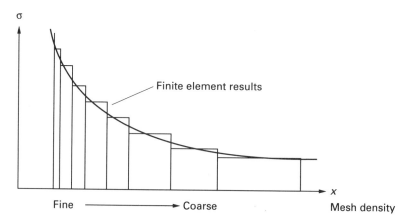

Figure 5.12 Approximation of a stress distribution using different sized elements and mesh densities

5.5   Three-dimensional elasticity

The application of a three-dimensional simplex element closely follows the two-dimensional implementation. The displacement components are defined in the usual way, and are stored most compactly as

$$\{u\} = [N]\{U\}$$

where $[N]$ is specified in Eq. 4.33.

The general form of the shape functions is given by Eq. 4.15 as

$$N_\beta = (a_\beta + b_\beta x + c_\beta y + d_\beta z)/6V \qquad \beta = i,j,k,l$$

The $[B]$ matrix relates the strains experienced in an element to the nodal displacements, and for a three-dimensional analysis we require all the six strain components as previously defined in Eq. 5.9. Consider for example the strain ε_x in the x direction. The variation of the displacement in the x direction is given by

$$u = N_i u_i + N_j u_j + N_k u_k + N_l u_l$$

Therefore the strain ε_x is

$$\varepsilon_x = \frac{\partial u}{\partial x} = \frac{\partial}{\partial x}(N_i u_i + N_j u_j + N_k u_k + N_l u_l)$$

$$= (b_i u_i + b_j u_j + b_k u_k + b_l u_l)/6V$$

Since all the terms in this equation are constants, the strain ε_x is constant for the element, as expected.

Similar derivations can be performed for the other two direct strains and the shear strains. They can all be summarized as follows:

$$
\begin{Bmatrix} \varepsilon_x \\ \varepsilon_y \\ \varepsilon_z \\ \gamma_{xy} \\ \gamma_{yz} \\ \gamma_{zx} \end{Bmatrix} = \frac{1}{6V}
\begin{bmatrix}
b_i & 0 & 0 & b_j & 0 & 0 & b_k & 0 & 0 & b_l & 0 & 0 \\
0 & c_i & 0 & 0 & c_j & 0 & 0 & c_k & 0 & 0 & c_l & 0 \\
0 & 0 & d_i & 0 & 0 & d_j & 0 & 0 & d_k & 0 & 0 & d_l \\
c_i & b_i & 0 & c_j & b_j & 0 & c_k & b_k & 0 & c_l & b_l & 0 \\
0 & d_i & c_i & 0 & d_j & c_j & 0 & d_k & c_k & 0 & d_l & c_l \\
d_i & 0 & b_i & d_j & 0 & b_j & d_k & 0 & b_k & d_l & 0 & b_l
\end{bmatrix}
\begin{Bmatrix} u_i \\ v_i \\ w_i \\ . \\ . \\ . \\ w_l \end{Bmatrix}
$$

$$= [B]\{U\} \tag{5.74}$$

The material property matrix $[D]$ for an isotropic material is

$$[D] = \frac{E}{(1+v)(1-2v)} \begin{bmatrix} 1-v & v & v & 0 & 0 & 0 \\ v & 1-v & v & 0 & 0 & 0 \\ v & v & 1-v & 0 & 0 & 0 \\ 0 & 0 & 0 & \dfrac{1-2v}{2} & 0 & 0 \\ 0 & 0 & 0 & 0 & \dfrac{1-2v}{2} & 0 \\ 0 & 0 & 0 & 0 & 0 & \dfrac{1-2v}{2} \end{bmatrix} \quad [5.75]$$

The stiffness matrix can then be calculated from

$$[k] = \int_V [B]^{\mathrm{T}}[D][B]\mathrm{d}V = [B]^{\mathrm{T}}[D][B]V \qquad [5.76]$$

The integration is trivial since both the $[B]$ and $[D]$ matrices only contain constants and can be removed from under the integral. Since the $[B]$ matrix is of size $[12 \times 6]$ and the $[D]$ matrix is $[6 \times 6]$, the stiffness matrix is a $[12 \times 12]$ square matrix, which corresponds to the twelve degrees of freedom of the element.

The calculation of the force vector terms proceeds in a similar manner to the two-dimensional analysis, but with extra coefficients leading to a final vector with twelve components. The initial strain vector is

$$\{\varepsilon_0\} = \alpha \Delta T \begin{Bmatrix} 1 \\ 1 \\ 1 \\ 0 \\ 0 \\ 0 \end{Bmatrix} \qquad [5.77]$$

Therefore, with a pressure applied to face ijk of the element, for example, the final general force vector term will be

$$\{F\} = \frac{vE\alpha\Delta T}{(1-2v)}[B]^{\mathrm{T}} \begin{Bmatrix} 1 \\ 1 \\ 1 \\ 0 \\ 0 \\ 0 \end{Bmatrix} + \frac{V}{4} \begin{Bmatrix} \mathbb{X} \\ \mathbb{Y} \\ \mathbb{Z} \\ . \\ . \\ \mathbb{Z} \end{Bmatrix} + \frac{S_{ijk}}{3} \begin{Bmatrix} p_x \\ p_y \\ p_z \\ . \\ . \\ 0 \end{Bmatrix} + \{P\} \qquad [5.78]$$

$$\quad\text{thermal}\qquad\qquad\text{body}\qquad\quad\text{pressure}\qquad\text{nodal}$$
$$\quad\text{expansion}\qquad\qquad\text{forces}\qquad\text{on side }ijk\qquad\text{forces}$$

5.6   Axisymmetric elasticity

A simple axisymmetric element is introduced in Section 4.6. It is similar to the two-dimensional element, except that it is used in the r–z plane as shown in Fig. 5.13. The stress and strain components for the element are

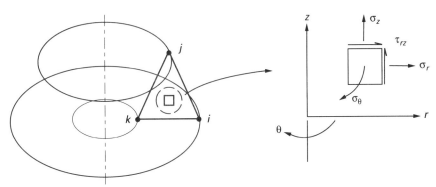

Figure 5.13 Basic axisymmetric element and stress components

$$\{\sigma\}^{\mathrm{T}} = [\sigma_r \ \sigma_\theta \ \sigma_z \ \tau_{rz}]$$
$$\{\varepsilon\}^{\mathrm{T}} = [\varepsilon_r \ \varepsilon_\theta \ \varepsilon_z \ \gamma_{rz}]$$ [5.79]

where the strains are defined as follows, with u and w being the displacements in the r and z directions respectively:

$$\varepsilon_r = \frac{\partial u}{\partial r} \qquad \varepsilon_\theta = \frac{u}{r} \qquad \varepsilon_z = \frac{\partial w}{\partial z} \qquad \gamma_{rz} = \frac{\partial u}{\partial z} + \frac{\partial w}{\partial r}$$ [5.80]

Note that the strain ε_θ in the circumferential direction is unusual since it is calculated from the actual value of the displacement rather than its derivation. If the $[B]$ matrix is calculated in the usual way, it is therefore found to be

$$[B] = \frac{1}{2A} \begin{bmatrix} b_i & 0 & b_j & 0 & b_k & 0 \\ \dfrac{2AN_i}{r} & 0 & \dfrac{2AN_j}{r} & 0 & \dfrac{2AN_k}{r} & 0 \\ 0 & c_i & 0 & c_j & 0 & c_k \\ c_i & b_i & c_j & b_j & c_k & b_k \end{bmatrix}$$ [5.81]

Normally with a linear element the differentiation of the displacement equations leads to a constant value, but calculation of the circumferential strain results in the matrix containing variables for the first time.

The $[D]$ matrix linking the stresses and strains is

$$[D] = \frac{E}{(1+\nu)(1-2\nu)} \begin{bmatrix} 1-\nu & \nu & \nu & 0 \\ \nu & 1-\nu & \nu & 0 \\ \nu & \nu & 1-\nu & 0 \\ 0 & 0 & 0 & \dfrac{1-2\nu}{2} \end{bmatrix}$$ [5.82]

The fact that ε_θ is not constant in an element means not only that the corresponding stress σ_θ varies through the element, but also that the other direct stresses σ_r and σ_z vary because they too are functions of ε_θ.

Since the $[B]$ matrix contains terms which are functions of the coordinates, evaluation of the stiffness matrix is no longer a trivial calculation as in the previous sections. The usual way that this is overcome is to evaluate the matrix for the centroid of the element, which has coordinates $\underline{r}$ and $\underline{z}$. This results in an approximate value for the $[B]$ matrix, denoted $[\underline{B}]$. The evaluation of the stiffness matrix then gives

$$[k] = [B]^{\mathrm{T}}[D][B]\bigg|_V \mathrm{d}V = [\underline{B}]^{\mathrm{T}}[D][\underline{B}]2\pi \underline{r} A \qquad [5.83]$$

This approximation is acceptable if the element is not being used to model a region with a high stress gradient.

Similar problems arise with the force vector terms, and again centroidal values are used to approximate the solution where the $[B]$ matrix is involved. For the thermal strain vector, any initial strain will be

$$\{\varepsilon_0\} = \alpha \Delta T \begin{Bmatrix} 1 \\ 1 \\ 1 \\ 0 \end{Bmatrix} \qquad [5.84]$$

Therefore the force vector term due to this strain is calculated from

$$\int_V [B]^{\mathrm{T}}[D]\{\varepsilon_0\}\mathrm{d}V = \frac{\alpha E \Delta T}{(1 - 2v)} [\underline{B}]^{\mathrm{T}} \begin{Bmatrix} 1 \\ 1 \\ 1 \\ 0 \end{Bmatrix} 2\pi \underline{r} A \qquad [5.85]$$

The body force and pressure terms can be evaluated precisely by the use of natural coordinates, together with the relationship linking the global and local coordinates, as introduced in Section 4.4 and in particular Eq. 4.24. For an axisymmetric element this means that a radial distance can be written as

$$r = L_1 r_i + L_2 r_j + L_3 r_k \qquad [5.86]$$

where r_i, r_j and r_k are the radial coordinates of the three nodes.

The body force vector is found from

$$\int_V [N]^{\mathrm{T}} \begin{Bmatrix} \mathbb{R} \\ \mathbb{Z} \end{Bmatrix} \mathrm{d}V = \int_V [N]^{\mathrm{T}} \begin{Bmatrix} \mathbb{R} \\ \mathbb{Z} \end{Bmatrix} 2\pi r \, \mathrm{d}A \qquad [5.87]$$

$\mathbb{R}$ and $\mathbb{Z}$ are the body forces per unit volume in the radial and axial directions respectively. Rather than using the centroid values for r in the above equation, the definition in Eq. 5.86 is used. Also, the shape functions are identical to the natural coordinates. Hence the calculation proceeds as follows:

$$\int_V \begin{bmatrix} rN_i & 0 \\ 0 & rN_i \\ rN_j & 0 \\ 0 & rN_j \\ rN_k & 0 \\ 0 & rN_k \end{bmatrix} \begin{Bmatrix} \mathbb{R} \\ \mathbb{Z} \end{Bmatrix} 2\pi \, \mathrm{d}A$$

$$= \int_V \begin{Bmatrix} (L_1^2 r_i + L_1 L_2 r_j + L_1 L_3 r_k)\mathbb{R} \\ (L_1^2 r_i + L_1 L_2 r_j + L_1 L_3 r_k)\mathbb{Z} \\ (L_2 L_1 r_i + L_2^2 r_j + L_2 L_3 r_k)\mathbb{R} \\ (L_2 L_1 r_i + L_2^2 r_j + L_2 L_3 r_k)\mathbb{Z} \\ (L_3 L_1 r_i + L_3 L_2 r_j + L_3^2 r_k)\mathbb{R} \\ (L_3 L_1 r_i + L_3 L_2 r_j + L_3^2 r_k)\mathbb{Z} \end{Bmatrix} 2\pi \, \mathrm{d}A$$

Using the factorial integration formula in the usual way, and specifically Eq. 4.26, this finally gives

$$\frac{\pi A}{6} \begin{Bmatrix} (2r_i + r_j + r_k)\mathbb{R} \\ (2r_i + r_j + r_k)\mathbb{Z} \\ (r_i + 2r_j + r_k)\mathbb{R} \\ (r_i + 2r_j + r_k)\mathbb{Z} \\ (r_i + r_j + 2r_k)\mathbb{R} \\ (r_i + r_j + 2r_k)\mathbb{Z} \end{Bmatrix} \qquad [5.88]$$

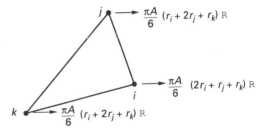

Figure 5.14 Distribution of a body force in an axisymmetric element

Note that this does not distribute the body forces equally between the three nodes. The node furthest from the centre of rotation (and largest value of r) will receive the largest share of the force, as illustrated in Fig. 5.14.

When surface pressures are considered, the total area of the element face as the element is rotated about the z axis must be included. Thus if the length of side ij in Fig. 5.13 is H_{ij}, the total area to be considered is

$2\pi r H_{ij}$. The force vector term arising from the pressure is then

$$\int_S [N]^\mathrm{T} \begin{Bmatrix} p_r \\ p_z \end{Bmatrix} \mathrm{d}S = \int_{H_{ij}} \begin{Bmatrix} p_r \\ p_z \end{Bmatrix} 2\pi r \, \mathrm{d}H$$

$$= \int_{H_{ij}} \begin{bmatrix} N_i & 0 \\ 0 & N_i \\ N_j & 0 \\ 0 & N_j \\ 0 & 0 \\ 0 & 0 \end{bmatrix} \begin{Bmatrix} p_r \\ p_z \end{Bmatrix} 2\pi r \, \mathrm{d}H$$

If r is replaced by Eq. 5.86 and natural coordinates are used in a similar way to the body force integral, the final pressure force term is found to be

$$\frac{\pi H_{ij}}{3} \begin{Bmatrix} (2r_i + r_j)p_r \\ (2r_i + r_j)p_z \\ (r_i + 2r_j)p_r \\ (r_i + 2r_j)p_z \\ 0 \\ 0 \end{Bmatrix} \qquad [5.89]$$

Thus the nodal forces which are used to replace the pressure loads will not necessarily be equal; again the node further from the axis of rotation is allocated the larger share. The only case where they are equal is when the element side is vertical. For this problem $r_i = r_j = r$, and the force vector is

$$\pi r H_{ij} \begin{Bmatrix} p_r \\ p_z \\ p_r \\ p_z \\ 0 \\ 0 \end{Bmatrix} \qquad [5.90]$$

In summary, the force vector equations for an axisymmetric element experiencing an initial strain, body forces, a pressure load on side ij and general nodal forces are as follows:

$$\frac{\alpha E \Delta T}{(1 - 2\nu)} [B]^T \begin{Bmatrix} 1 \\ 1 \\ 1 \\ 0 \end{Bmatrix} 2\pi r A + \frac{\pi A}{6} \begin{Bmatrix} (2r_i + r_j + r_k)\mathbb{R} \\ (2r_i + r_j + r_k)\mathbb{Z} \\ (r_i + 2r_j + r_k)\mathbb{R} \\ (r_i + 2r_j + r_k)\mathbb{Z} \\ (r_i + r_j + 2r_k)\mathbb{R} \\ (r_i + r_j + 2r_k)\mathbb{Z} \end{Bmatrix}$$

thermal
expansion   body
forces

$$+ \frac{\pi H_{ij}}{3} \begin{Bmatrix} (2r_i + r_j)p_r \\ (2r_i + r_j)p_z \\ (r_i + 2r_j)p_r \\ (r_i + 2r_j)p_z \\ 0 \\ 0 \end{Bmatrix} + \{P\} \qquad\qquad [5.91]$$

pressure
on side ij nodal
forces

Example 5.5: axisymmetric stress analysis

The mesh shown in Fig. 5.15 is part of an axisymmetric finite element model to examine the stresses in a cylinder with an internal pressure of 5 MPa and an applied axial pressure of -10 MPa. Examine how these pressure loads are included in the finite element calculations, and in particular what nodal loads are used to represent them.

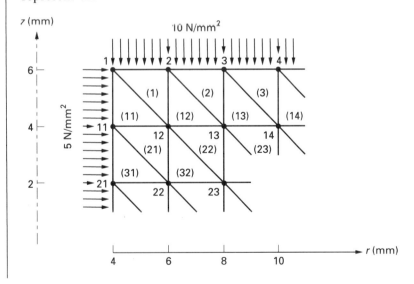

Figure 5.15

Firstly, considering element (1), and assuming nodes 2 and 1 are the i and j nodes respectively of the general element, then Eq. 5.89 is used to calculate the force vector with $p_r = 0$. Substituting in the radial coordinates of the nodes and the length of the side H_{ij} gives

$$\frac{\pi 2}{3} \begin{Bmatrix} 0 \\ -10 \times (2 \times 6 + 4) \\ 0 \\ -10 \times (6 + 2 \times 4) \\ 0 \\ 0 \end{Bmatrix} = \begin{Bmatrix} 0 \\ -335 \\ 0 \\ -293 \\ 0 \\ 0 \end{Bmatrix} \qquad [5.92]$$

Elements (2) and (3) similarly only experience an axial pressure, and calculation of the corresponding force vector terms gives

$$\{F^{(2)}\} = \begin{Bmatrix} 0 \\ -461 \\ 0 \\ -419 \\ 0 \\ 0 \end{Bmatrix} \qquad \{F^{(3)}\} = \begin{Bmatrix} 0 \\ -586 \\ 0 \\ -545 \\ 0 \\ 0 \end{Bmatrix} \qquad [5.93]$$

For the internal pressure acting on the vertical side of element (11), assuming the nodes are $i = 1$ and $j = 11$, and noting that $p_r = 5$ MPa with $p_z = 0$, the force vector is

$$\frac{\pi 2}{3} \begin{Bmatrix} (3 \times 4) \times 5 \\ 0 \\ (3 \times 4) \times 5 \\ 0 \\ 0 \\ 0 \end{Bmatrix} = \begin{Bmatrix} 126 \\ 0 \\ 126 \\ 0 \\ 0 \\ 0 \end{Bmatrix} \qquad [5.94]$$

This time the nodal forces are the same, and similar vectors are calculated for all the other elements on the inside of the cylinder, since their dimensions and orientations are the same as those of element (1). Therefore the pressure loading detailed above is actually applied in the finite element model as shown in Fig. 5.16.

The application of the pressure load on the horizontal face is by no means obvious, although one would expect the forces to increase further away from the axis of symmetry, because the effective surface area over which the pressure acts is larger. For element (2), for example, the total force carried by the element is

$$10\pi(8^2 - 6^2) = 880 \text{ N}$$

The 880 N equals the sum of the required nodal forces (461 + 419 N) as previously calculated in Eq. 5.93.

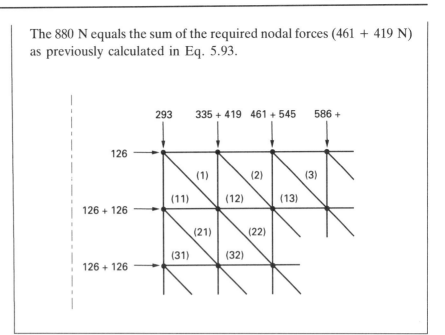

Figure 5.16

5.7 Conclusions

This chapter introduces the general finite element equations for elasticity problems. The equations are applied to one-, two- and three-dimensional and axisymmetric analyses with simplex elements. These elements assume a linear variation of displacement, and consequently lead to constant stress and strain fields (with the exception of the axisymmetric implementation). Since they are constant stress elements, they invariably require a fine mesh when used to model real engineering problems.

In practice, if the computer resources are sufficient, then higher-order elements are preferred because they allow linear and possibly quadratic stress variations. The simplex elements, however, are important and are ideal for introducing the basic concepts of the finite element method. The more sophisticated elements require numerical integration, and are not so suitable for the type of examples used in this chapter. Chapter 8 discusses how higher-order elements are implemented. They use the same governing equations, and result in similar stiffness matrix and force vector terms, but include more coefficients because of the larger number of degrees of freedom in each element.

Problems

5.1   Analyse the bars shown in Fig. 5.17(a)–(f) using one-dimensional elements, and compare the results with those calculated theoretically where possible. Assume $E = 200$ GPa and $\alpha = 2 \times 10^{-6}$ per °C in all cases.

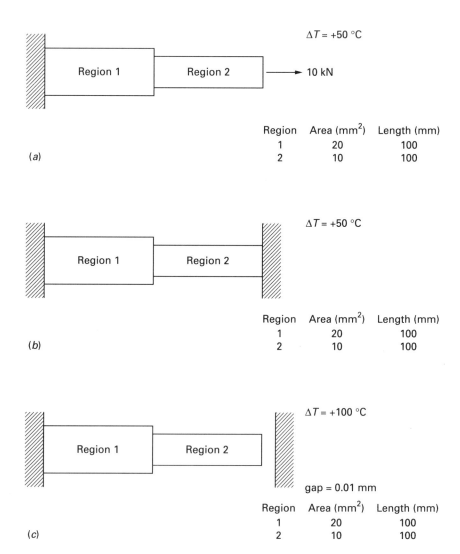

Region	Area (mm²)	Length (mm)
1	20	100
2	10	100

(a)

Region	Area (mm²)	Length (mm)
1	20	100
2	10	100

(b)

Region	Area (mm²)	Length (mm)
1	20	100
2	10	100

(c)

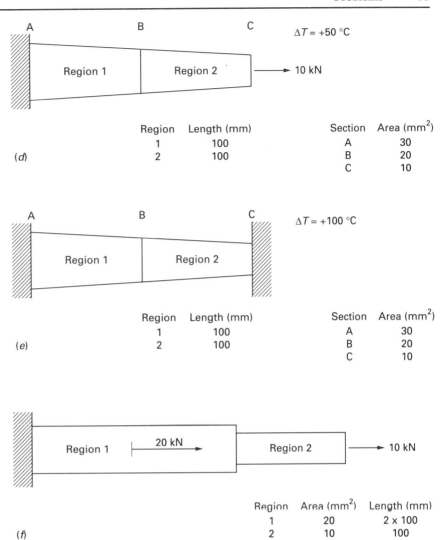

Figure 5.17

5.2   Analyse the trusses shown in Fig. 5.18(a)–(d) using two-dimensional truss elements, and compare the axial forces predicted with those calculated theoretically. Assume all members are manufactured from steel ($E = 200 \times 10^3$ MPa).

5.3   By considering Hooke's law, confirm that the $[D]$ property matrices for plane stress and plane strain approximations are given by Eqs 5.56 and 5.57.

5.4   Calculate the term required in the force vector to account for thermal strain in a plane strain triangular simplex element.

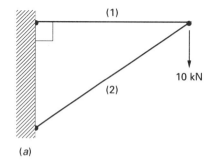

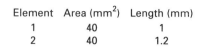

Element	Area (mm²)	Length (mm)
1	40	1
2	40	1.2

(a)

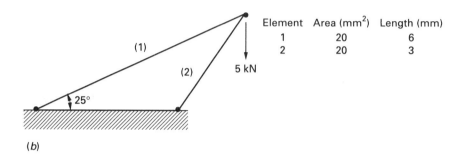

Element	Area (mm²)	Length (mm)
1	20	6
2	20	3

(b)

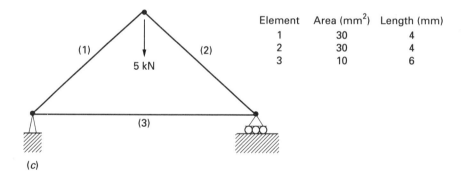

Element	Area (mm²)	Length (mm)
1	30	4
2	30	4
3	10	6

(c)

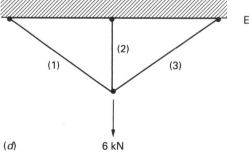

Element	Area (mm²)	Length (mm)
1	20	3.5
2	20	2
3	20	3.5

Figure 5.18

(d)   6 kN

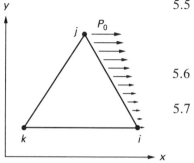

Figure 5.19

5.5 Derive the force vector term required to account for the linearly varying pressure load shown in Fig. 5.19. [Hint: express the pressure distribution as a function of the nodal values using the shape functions.]

5.6 Calculate the finite element equations for the simplex elements shown in Fig. 5.20(a)–(e).

5.7 The calculated displacements $(10^{-3}\,\text{mm})$ of the elements in Problem 5.6 are shown in the following table. What are the resulting element stresses?

Element	Node i		Node j		Node k	
	u	v	u	v	u	v
(a)	0	0	1	2	−3	2
(b)	−5	0	2	2	1	−1
(c)	1	2	6	−3	4	2
(d)	1	−3	2	4	−1	6
(e)	2	4	3	−1	2	5

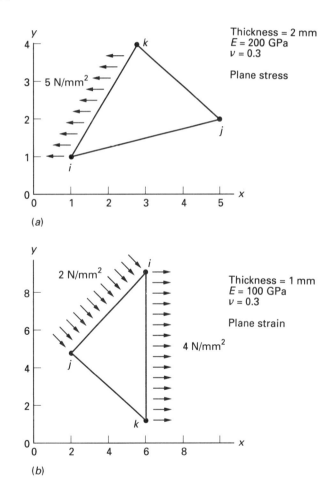

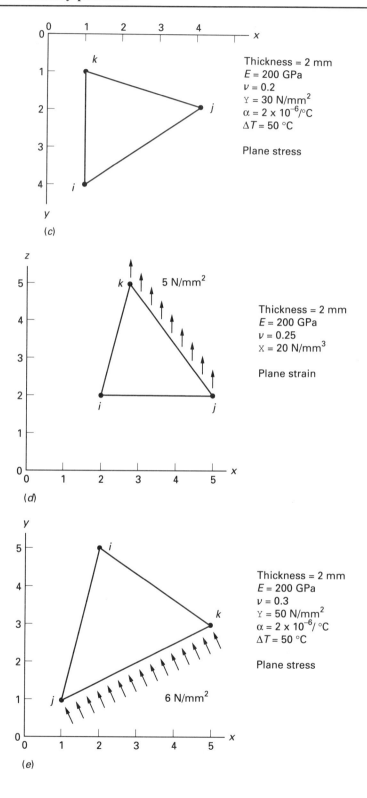

Thickness = 2 mm
E = 200 GPa
ν = 0.2
Y = 30 N/mm^2
α = 2 x 10^{-6}/°C
ΔT = 50 °C

Plane stress

(c)

5 N/mm^2

Thickness = 2 mm
E = 200 GPa
ν = 0.25
X = 20 N/mm^3

Plane strain

(d)

Thickness = 2 mm
E = 200 GPa
ν = 0.3
Y = 50 N/mm^2
α = 2 x 10^{-6}/ °C
ΔT = 50 °C

Plane stress

6 N/mm^2

(e)

Figure 5.20

5.8   Prove that for an axisymmetric element with a pressure load on side ik, the relevant force vector term is

$$\frac{\pi H_{ik}}{3}\begin{Bmatrix} (2r_i + r_k)p_r \\ (2r_i + r_k)p_z \\ 0 \\ 0 \\ (r_i + 2r_k)p_r \\ (r_i + 2r_k)p_z \end{Bmatrix}$$

5.9   Calculate the finite element equations for the axisymmetric simplex elements shown in Fig. 5.21(a)–(c).

5.10 The calculated displacements $(10^{-3}\,\text{mm})$ of the elements in Problem 5.9 are shown in the following table. What are the resulting element stresses?

Element	Node i		Node j		Node k	
	r	z	r	z	r	z
(a)	1	3	−2	5	2	2
(b)	2	−4	6	0	0	1
(c)	5	0	1	−3	1	2

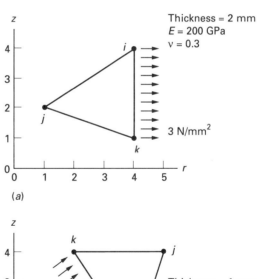

(a)

Thickness = 2 mm
$E = 200$ GPa
$v = 0.3$

3 N/mm^2

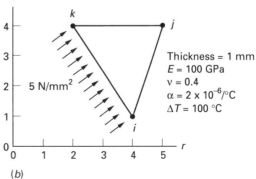

(b)

Thickness = 1 mm
$E = 100$ GPa
$v = 0.4$
$\alpha = 2 \times 10^{-6}/°C$
$\Delta T = 100\ °C$

5 N/mm^2

Figure 5.21

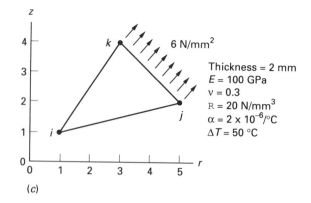

(c)

6 Formulation of the element characteristic matrices and vectors for field problems

6.1 Introduction

Many important problems in engineering are governed by the field equation, and can be conveniently and easily analysed by the finite element method. The most common of these are heat transfer, torsion of shafts, irrotational flow, groundwater seepage, electrostatic and magneto-static fields and fluid film lubrication. The equation that governs all these processes can be stated as

$$\frac{\partial}{\partial x}\left(K_x\frac{\partial \phi}{\partial x}\right) + \frac{\partial}{\partial y}\left(K_y\frac{\partial \phi}{\partial y}\right) + \frac{\partial}{\partial z}\left(K_z\frac{\partial \phi}{\partial z}\right) + Q = 0 \qquad [6.1]$$

Since this equation is common to the different areas, a general approach can be used to derive the governing element equations, which can then be adjusted to represent the particular problem under consideration.

The boundary conditions associated with Eq. 6.1 in their most general forms are

$$\phi = \phi_1 \qquad [6.2]$$

on surface S_1, and

$$K_x\frac{\partial \phi}{\partial x}l_x + K_y\frac{\partial \phi}{\partial y}l_y + K_z\frac{\partial \phi}{\partial z}l_z + q = 0 \qquad [6.3]$$

$$K_x\frac{\partial \phi}{\partial x}l_x + K_y\frac{\partial \phi}{\partial y}l_y + K_z\frac{\partial \phi}{\partial z}l_z + h(\phi - \phi_\infty) = 0 \qquad [6.4]$$

on surfaces S_2 and S_3 respectively, where S_1, S_2 and S_3 form the complete boundary of the region under consideration. The term q and the coefficients K_x, K_y and K_z may be functions of the coordinates x, y and z, but are assumed to be independent of the variable ϕ. l_x, l_y and l_z are the direction cosines of the outward normal to the boundary.

The field equation is probably most familiar from heat transfer problems, where the unknown variable is temperature. K_x, K_y and K_z represent the thermal conductivities of the material in three directions, Q is an internal heat source or sink, h is the convection coefficient at the surface and q is an applied heat flux acting over the surface.

For torsion problems, consider a two-dimensional situation where $K_x = K_y = 1$ and $Q = 2G\theta_1$, with the boundary condition of ϕ equalling a constant (usually zero) around the boundary, so that

$$\frac{\partial^2\phi}{\partial x^2} + \frac{\partial^2\phi}{\partial y^2} = -2G\phi_1 \qquad [6.5]$$

Here ϕ is Prandtl's stress function, G is the shear modulus of the material and θ_1 the twist per unit length of the bar. The shear stresses in the bar are calculated by the derivatives of ϕ in the x and y directions, and the torque by the volume under the ϕ surface.

When irrotational flow is considered, $K_x = K_y = 1$ and $Q = 0$, so that the resulting two-dimensional equation is

$$\frac{\partial^2\phi}{\partial x^2} + \frac{\partial^2\phi}{\partial y^2} = 0 \qquad [6.6]$$

which can be formulated in terms of a velocity potential function, or a stream function.

For seepage problems, the general equation remains as

$$K_x\frac{\partial^2\phi}{\partial x^2} + K_y\frac{\partial^2\phi}{\partial y^2} + Q = 0 \qquad [6.7]$$

where the permeability of the soil is represented by coefficients K_x and K_y, Q is the fluid source (or sink), and the variable ϕ is the fluid potential or piezometric head.

If the equation is applied to electrostatic fields, the K coefficients are used to model the permittivities in three directions, the Q term represents any internal current sources, and the field variable ϕ the electric force field intensity.

For magnetostatics, the finite element method can be used to solve for the magnetomotive force if K_x, K_y and K_z equal the magnetic permeabilities and Q is any internal magnetic field source.

Finally, for fluid film lubrication problems, the pressure acting in the lubricant can be modelled, where K_x and K_y are functions of film thickness and viscosity, and Q represents any internal source of lubrication.

Clearly, the range of problems that are covered by the field equation is diverse, and a common analysis approach using the finite element method appears attractive. The general element equations governing field problems are now developed, and are applied in detail to heat transfer problems and the torsion of shafts.

6.2   Formulation procedures

The basic finite element equations for field problems can be derived in a number of ways, but the most widely used methods are the variational approach and the method of weighted residuals.

The variational formulation relies on the minimization of a functional. In structural and solid mechanics the functional turns out to be the potential energy of the system, but for field problems a functional must be derived from the governing differential equations and the boundary conditions. Once the functional is known, the formulation of the finite element equations for field problems proceeds in the same way as the solid mechanics derivation, with the equations taking up a similar format.

The majority of problems for analysis by the finite element method can be solved using a variational formulation, but where a functional cannot be derived another approach needs to be adopted. The weighted residual method starts directly from the differential equation of the problem, and relies on minimization of the error (or residual) incurred by the use of an assumed distribution of the unknown variable.

Both the variational formulation and the weighted residual method (specifically Galerkin's method) are now used to derive the governing equations for field problems.

6.2.1   The variational formulation

The most straightforward way of deriving the governing equations for field problems is to use a variational formulation, by minimizing a functional. (A functional can be defined as a function of several other functions.) For field problems the functional is derived from the governing differential equation Eq. 6.1 and the boundary conditions Eqs 6.2 to 6.4. It possesses the property that any function which makes it a minimum also satisfies the governing differential equation and the boundary conditions.

The technique by which functionals are derived is part of a branch of mathematics called the calculus of variations, and is not repeated in this book since it is well covered elsewhere in the literature. The important result from the calculus of variations is that the minimization of the functional

$$I = \int_V \frac{1}{2} \left[K_x \left(\frac{\partial \phi}{\partial x} \right)^2 + K_y \left(\frac{\partial \phi}{\partial y} \right)^2 + K_z \left(\frac{\partial \phi}{\partial z} \right)^2 - 2Q\phi \right] \mathrm{d}V$$
$$+ \int_{S_2} q\phi \, \mathrm{d}S + \int_{S_3} \frac{h}{2}(\phi - \phi_\infty)^2 \mathrm{d}S \quad [6.8]$$

requires that the governing differential equation Eq. 6.1 and the boundary conditions Eqs 6.2 to 6.4 are all satisfied.

To put the functional I into a more manageable format, the following substitutions are made for the vector of gradients of the field variable

$$\{g\}^{\mathrm{T}} = \left[\begin{array}{ccc} \dfrac{\partial \phi}{\partial x} & \dfrac{\partial \phi}{\partial y} & \dfrac{\partial \phi}{\partial z} \end{array} \right] \tag{6.9}$$

and for the matrix of material properties

$$[D] = \left[\begin{array}{ccc} K_x & 0 & 0 \\ 0 & K_y & 0 \\ 0 & 0 & K_z \end{array} \right] \tag{6.10}$$

Therefore Eq. 6.8 becomes

$$I = \int_V \frac{1}{2} [\{g\}^{\mathrm{T}} [D]\{g\} - 2Q\phi] \mathrm{d}V + \int_{S_2} q\phi \, \mathrm{d}S$$

$$+ \int_{S_3} \frac{h}{2} (\phi^2 - 2\phi\phi_\infty + \phi_\infty^2) \mathrm{d}S \tag{6.11}$$

Now since the variable is to be modelled in a discrete way by a finite element representation rather than continuously, the functional I must be expressed as the E elemental quantities $I^{(e)}$, so that

$$I = \sum_{e=1}^{E} I^{(e)} \tag{6.12}$$

where

$$I^{(e)} = \int_{V^{(e)}} \frac{1}{2} [\{g^{(e)}\}^{\mathrm{T}} [D^{(e)}]\{g^{(e)}\} - 2Q^{(e)}\phi] \mathrm{d}V + \int_{S_2^{(e)}} q^{(e)}\phi \, \mathrm{d}S$$

$$+ \int_{S_3^{(e)}} \frac{h}{2} (\phi^2 - 2\phi\phi_\infty + \phi_\infty^2) \mathrm{d}S \tag{6.13}$$

Also the behaviour of the function ϕ within each element can be expressed in the usual way as

$$\phi = [N^{(e)}]\{\Phi^{(e)}\} \tag{6.14}$$

where $[N^{(e)}]$ is the matrix of shape functions for element (e) and $\{\Phi^{(e)}\}$ is the vector of unknown nodal variables for element (e). Therefore

$$\{g^{(e)}\} = [B^{(e)}]\{\Phi^{(e)}\} \tag{6.15}$$

with $[B^{(e)}]$ calculated by the proper differentiation of $[N^{(e)}]$.

Hence using Eqs 6.14 and 6.15, the discretized form of the functional for one element becomes

$$I^{(e)} = \int_{V^{(e)}} \frac{1}{2} \{\Phi^{(e)}\}^{\mathrm{T}} [B^{(e)}]^{\mathrm{T}} [D^{(e)}] [B^{(e)}] \{\Phi^{(e)}\} \mathrm{d}V$$

$$- \int_{V^{(e)}} Q^{(e)} [N^{(e)}] \{\Phi^{(e)}\} \mathrm{d}V + \int_{S_2^{(e)}} q^{(e)} [N^{(e)}] \{\Phi^{(e)}\} \mathrm{d}S$$

$$+ \int_{S_3^{(e)}} \frac{h}{2} \{\Phi^{(e)}\}^{\mathrm{T}} [N^{(e)}]^{\mathrm{T}} [N^{(e)}] \{\Phi^{(e)}\} \ \mathrm{d}S - \int_{S_3^{(e)}} h\phi_\infty [N^{(e)}] \{\Phi^{(e)}\} \mathrm{d}S$$

$$+ \int_{S_3^{(e)}} \frac{h}{2} \phi_\infty^2 \ \mathrm{d}S \qquad\qquad [6.16]$$

The variational method requires that the functional is minimized, that is, values of the nodal variables must be determined which cause I to be a minimum. The usual approach of differentiating and setting the result equal to zero is employed. For the finite element method this means differentiating with respect to each of the unknown nodal variables. Thus if there are n degrees of freedom, n equations will be produced. If $\{\Phi_S\}$ is a vector of the unknown variables for the whole system, this can be expressed in terms of the previous equations as

$$\frac{\partial I}{\partial \{\Phi_s\}} = \frac{\partial}{\partial \{\Phi_s\}} \sum_{e=1}^{E} I^{(e)} = \sum_{e=1}^{E} \frac{\partial I^{(e)}}{\partial \{\Phi^{(e)}\}}$$

This minimization of the elemental quantity $I^{(e)}$ gives

$$\frac{\partial I^{(e)}}{\partial \{\Phi^{(e)}\}} = \left(\int_{V^{(e)}} [B^{(e)}]^{\mathrm{T}} [D^{(e)}] [B^{(e)}] \mathrm{d}V \right.$$

$$\left. + \int_{S_3^{(e)}} h[N^{(e)}]^{\mathrm{T}} [N^{(e)}] \mathrm{d}S \right) \{\Phi^{(e)}\} - \int_{V^{(e)}} Q^{(e)} [N^{(e)}]^{\mathrm{T}} \mathrm{d}V$$

$$+ \int_{S_2^{(e)}} q^{(e)} [N^{(e)}]^{\mathrm{T}} \mathrm{d}S - \int_{S_3^{(e)}} h\phi_\infty [N^{(e)}]^{\mathrm{T}} \mathrm{d}S \qquad [6.17]$$

This can be written as

$$\frac{\partial I^{(e)}}{\partial \{\Phi^{(e)}\}} = [k^{(e)}] \{\Phi^{(e)}\} - \{F^{(e)}\} \qquad\qquad [6.18]$$

where $[k^{(e)}]$ is the element stiffness matrix and $\{F^{(e)}\}$ is the element force vector.

For the whole system,

$$\frac{\partial I}{\partial \{\Phi_S\}} = \sum_{e=1}^{E} ([k^{(e)}] \{\Phi^{(e)}\} - \{F^{(e)}\}) = 0 \qquad\qquad [6.19]$$

and assembly of the element equations to obtain the system equations proceeds in the normal manner.

6.2.2   The weighted residual method

Consider a differential equation

$$D(\phi) - F = 0 \tag{6.20}$$

where D is a differential operator acting on an unknown function ϕ. For example, the equation might simply be

$$\frac{d^2\phi}{dx^2} - F = 0 \tag{6.21}$$

The weighted residual method involves substituting an approximate solution into the governing differential equation, and then working with the resulting error or residual.

For example, if an approximation $\phi(x)$ is used with

$$\phi(x) = \sum N_i\phi_i \qquad i = 1,2,\ldots,n \tag{6.22}$$

where ϕ_i are constants and N_i are independent functions of x, then Eq. 6.21 would yield a residual R, that is

$$\frac{d^2\phi(x)}{dx^2} - F = R \neq 0 \tag{6.23}$$

The residual is multiplied by a weighting function W (which is a function of x), and the integral of the product is then required to be zero. Therefore

$$\int_r W_i R \, dx = 0 \tag{6.24}$$

over the region r, where the number of weighting functions is equal to the number of coefficients in the assumed solution (Eq. 6.22).

Different weighting functions may be chosen, but the most widely used approach is known as Galerkin's method. This uses the same weighting functions that are used in the approximating equation (*i.e.* Eq. 6.22). Hence

$$\int_r N_i R \, dx = 0 \tag{6.25}$$

Equation 6.22 is the same form of equation as used previously for the interpolation function in the simplex elements in Chapter 4, and N_i are the shape functions.

There is one problem associated with Eq. 6.25. The highest-order derivative allowed in the integral has an order that is one greater than the order of continuity in the interpolation equations. For example, with a linear interpolation function (*i.e.* a simplex element) with continuity in ϕ, only a first-order derivative can appear in the integral. Fortunately the difficulty can be overcome by reducing the order of the derivative by integration by parts.

The concept behind the finite element method is of course to discretize the region or body under consideration, so the basic equation Eq. 6.25 must be converted into an elemental form. The number of weighting coefficients is equal to the number of nodes in the model, and therefore Eq. 6.25 is composed of n equations for a model with n nodes. Therefore, for a general body V,

$$\int_V [N]^T R \, dV = 0 \tag{6.26}$$

where $[N] = (N_1 \, N_2 \, N_3 \ldots N_n)$. But if the region is divided into E elements,

$$\sum_{e=1}^{E} \int_{V^{(e)}} [N^{(e)}]^T R^{(e)} \, dV = \sum_{e=1}^{E} \{G^{(e)}\} = 0 \tag{6.27}$$

The following use of the weighted residual method in the development of the governing equations of field problems should clarify the procedure.

Application to field problems

Working from the general field equation Eq. 6.1, the contribution from an element (e) is

$$\int_{V^{(e)}} [N^{(e)}]^T R^{(e)} \, dV$$
$$= \int_{V^{(e)}} [N^{(e)}]^T \left(K_x \frac{\partial^2 \phi}{\partial x^2} + K_y \frac{\partial^2 \phi}{\partial y^2} + K_z \frac{\partial^2 \phi}{\partial z^2} + Q \right) dV \tag{6.28}$$

where $[N^{(e)}]$ is the row vector containing the element shape functions. The second derivatives in Eq. 6.28 must be replaced by first derivatives using integration by parts. Consider

$$\int_{V^{(e)}} [N^{(e)}]^T K_x \frac{\partial^2 \phi}{\partial x^2} \, dV$$
$$= \int_{V^{(e)}} \left(K_x \frac{\partial}{\partial x} \left([N^{(e)}]^T \frac{\partial \phi}{\partial x} \right) - K_x \frac{\partial [N^{(e)}]^T}{\partial x} \frac{\partial \phi}{\partial x} \right) dV \tag{6.29}$$

The first term on the right hand side can be replaced by the use of Green's theorem to give

$$\int_{V^{(e)}} K_x \frac{\partial}{\partial x} \left([N^{(e)}]^T \frac{\partial \phi}{\partial x} \right) dV = \int_{S^{(e)}} K_x [N^{(e)}]^T \frac{\partial \phi}{\partial x} l_x \, dS \tag{6.30}$$

where $S^{(e)}$ is a surface of the element and l_x is the x direction cosine of the normal to the surface.

The other y and z second-order terms in the integral of Eq. 6.28 can also be expressed in the same way as the x term. The final set of equations from the process is then

$$\{G^{(e)}\} = - \int_{V^{(e)}} \left(K_x \frac{\partial [N^{(e)}]^{\mathrm{T}}}{\partial x} \frac{\partial \phi}{\partial x} + K_y \frac{\partial [N^{(e)}]^{\mathrm{T}}}{\partial y} \frac{\partial \phi}{\partial y} \right.$$

$$\left. + K_z \frac{\partial [N^{(e)}]^{\mathrm{T}}}{\partial z} \frac{\partial \phi}{\partial z} \right) \mathrm{d}V + \int_{S^{(e)}} [N^{(e)}]^{\mathrm{T}} \left(K_x \frac{\partial \phi}{\partial x} l_x + K_y \frac{\partial \phi}{\partial y} l_y \right.$$

$$\left. + K_z \frac{\partial \phi}{\partial z} l_z \right) \mathrm{d}S + \int_{V^{(e)}} Q^{(e)} [N^{(e)}]^{\mathrm{T}} \mathrm{d}V \qquad [6.31]$$

If the variation of ϕ across the element is expressed in the usual way as

$$\phi = [N^{(e)}]\{\Phi^{(e)}\} \qquad [6.32]$$

then the first term of Eq. 6.31 can be written as

$$- \left(\int_{V^{(e)}} \left(K_x \frac{\partial [N^{(e)}]^{\mathrm{T}}}{\partial x} \frac{\partial [N^{(e)}]}{\partial x} + K_y \frac{\partial [N^{(e)}]^{\mathrm{T}}}{\partial y} \frac{\partial [N^{(e)}]}{\partial y} \right. \right.$$

$$\left. \left. + K_z \frac{\partial [N^{(e)}]^{\mathrm{T}}}{\partial z} \frac{\partial [N^{(e)}]}{\partial z} \right) \mathrm{d}V \right) \{\Phi^{(e)}\} \qquad [6.33]$$

Now the gradient vector $\{g\}$ and the $[B]$ matrix are defined as

$$\{g^{(e)}\} = \begin{Bmatrix} \dfrac{\partial \phi}{\partial x} \\ \dfrac{\partial \phi}{\partial y} \\ \dfrac{\partial \phi}{\partial z} \end{Bmatrix} = \begin{Bmatrix} \dfrac{\partial [N^{(e)}]}{\partial x} \\ \dfrac{\partial [N^{(e)}]}{\partial y} \\ \dfrac{\partial [N^{(e)}]}{\partial z} \end{Bmatrix} \{\Phi^{(e)}\} = [B^{(e)}]\{\Phi^{(e)}\} \qquad [6.34]$$

Equation 6.33 becomes

$$- \left(\int_{V^{(e)}} [B^{(e)}]^{\mathrm{T}} [D^{(e)}][B^{(e)}] \mathrm{d}V \right) \{\Phi^{(e)}\} \qquad [6.35]$$

where the thermal conductivities are stored as

$$[D^{(e)}] = \begin{bmatrix} K_x & 0 & 0 \\ 0 & K_y & 0 \\ 0 & 0 & K_z \end{bmatrix} \qquad [6.36]$$

The possible boundary conditions for field problems are stated in Eqs 6.3 and 6.4. For surface S_2,

$$q = - K_x \frac{\partial \phi}{\partial x} l_x - K_y \frac{\partial \phi}{\partial y} l_y - K_z \frac{\partial \phi}{\partial z} l_z \qquad [6.37]$$

and for surface S_3,

$$h(\phi - \phi_\infty) = - K_x \frac{\partial \phi}{\partial x} l_x - K_y \frac{\partial \phi}{\partial y} l_y - K_z \frac{\partial \phi}{\partial z} l_z \qquad [6.38]$$

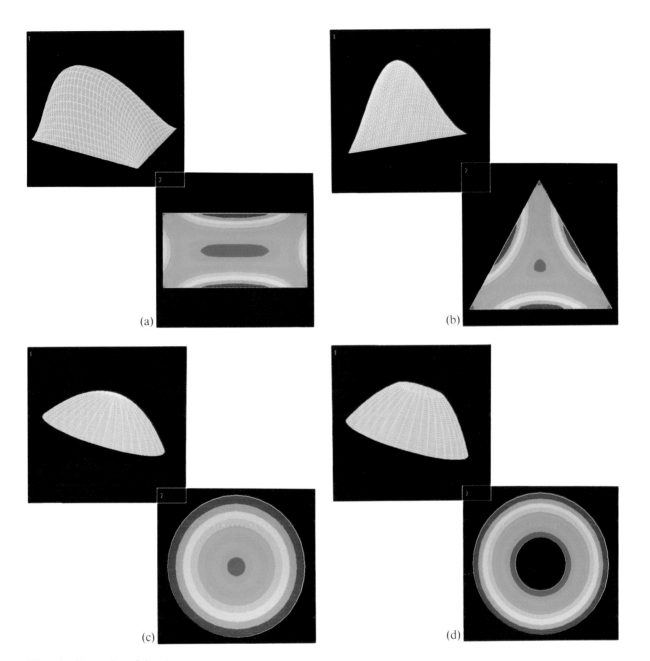

Plate 1 Examples of the ϕ surface and shear stress distribution in various cross-sections under torsion:
(a) square shaft,
(b) triangular shaft,
(c) circular shaft,
(d) hollow circular shaft (the minimum shear stress contour is dark-blue and the maximum is red)

Plate 2 The σ_x axial stress distribution in the lug of a pinned connection modelled in three different ways (see Fig. 9.7). (a) pin connected to lug with gap elements, (b) perfect pin to lug contact, (c) pin represented as rigid links. The run times for the three models were 213, 90 and 62 seconds respectively. The nine stress contours are equally spaced between −50 and 300 MPa (dark blue to red)

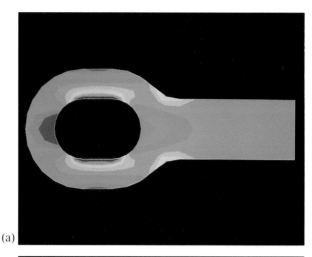

(a)

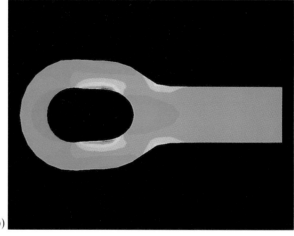

(b)

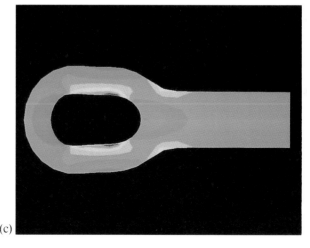

(c)

Plate 3 The deflected shapes and contours of vertical displacement (U_y) for three axially loaded bars modelled with different constraints overlaid on the original bar geometry (see Fig. 9.10). Only half the bar is represented because the problem is symmetric. The left-hand end lies on the line of symmetry, hence all the nodes on this face are constrained in the x-direction, and in addition: (a) the centre node is constrained in the y-direction – the correct constraint condition, (b) all nodes are constrained in the y-direction, (c) the bottom node is constrained in the y-direction. The fifteen displacement contours are equally spaced between -0.0075 and 0.015 mm (dark blue to red)

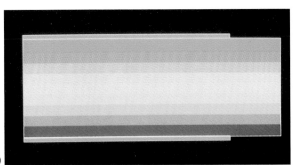

(a)

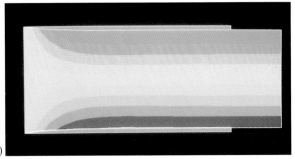

(b)

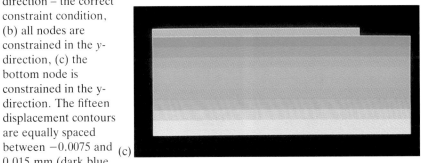

(c)

Plate 4 Maximum principal stress vector plot showing the load path around a hole, where the size and colour of the arrows reflects the magnitude of the stress; (dark-blue is the minimum, and red the maximum)

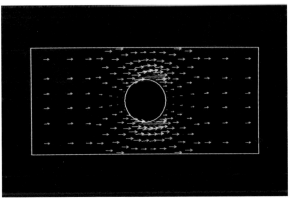

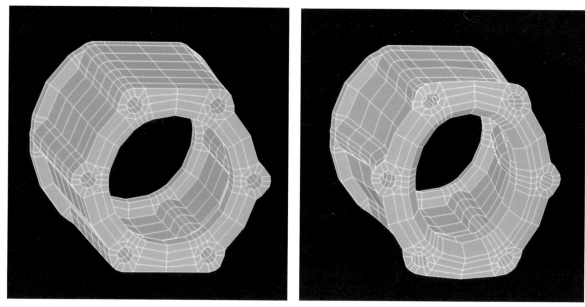

Plate 10 Finite element model
         and mode of vibration
         of a pump casing

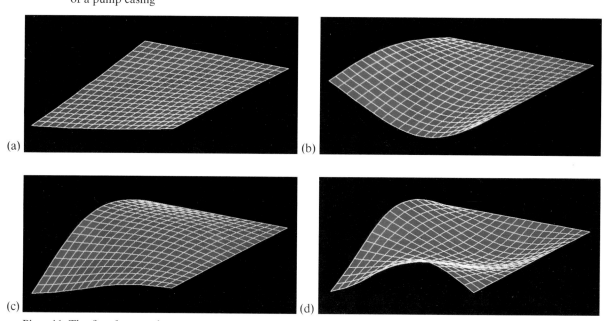

(a)

(b)

(c)

(d)

Plate 11 The first four modes
         of vibration of a
         square plate simply
         supported along two
         of its edges
         (frequencies of
         vibration are 120, 640,
         770 and 1470 Hz)

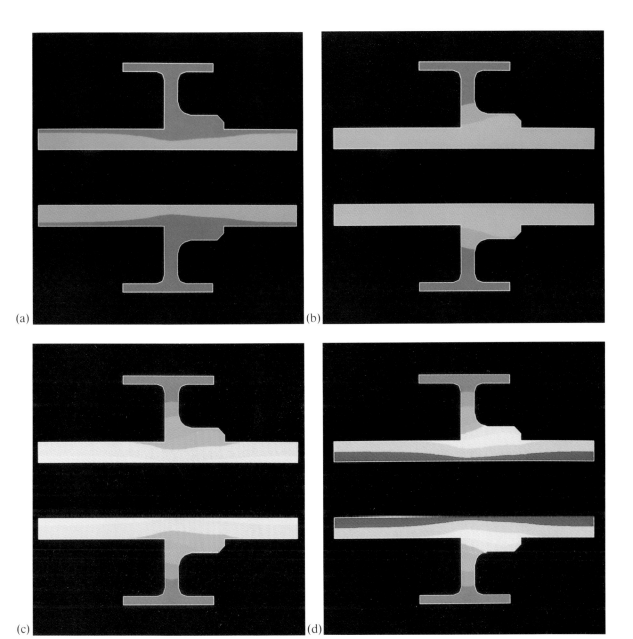

(a)

(b)

(c)

(d)

Plate 12 The temperature distribution in a pipe with supporting flange when a liquid at 200 °C is suddenly flushed along it. Plots are taken at 0.5, 2, 3 and 4 seconds, with temperature contours equally spaced between 20 °C and 200 °C (dark-blue to red)

Plate 13 Screen layout of the ANSYS user interface, allowing program commands and data to be entered through a series of menus and function pads using a mouse and/or keyboard. The menus are context sensitive, showing only the options that are valid for a particular analysis, and a sophisticated on-line help system is available for most aspects of the program and procedures (courtesy of Swanson Analysis Systems Inc.)

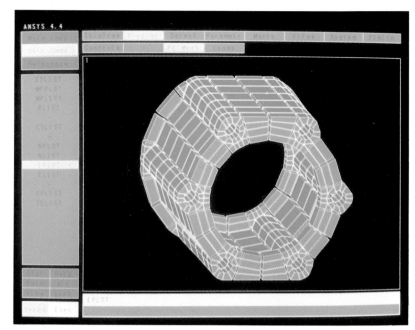

Plate 14 Finite element model of a turbine blade developed by automatic mesh generation (courtesy of Swanson Analysis Systems Inc.)

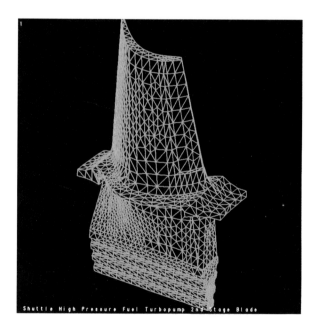

Thus using Eq. 6.32, the second term in Eq. 6.31 could be replaced by two terms for the surfaces S_2 and S_3, namely

$$- \int_{S_2^{(e)}} [N^{(e)}]^\mathrm{T} q^{(e)} \, \mathrm{d}S - \int_{S_3^{(e)}} [N^{(e)}]^\mathrm{T} h(\phi - \phi_\infty) \mathrm{d}S \qquad [6.39]$$

But if $\phi = [N^{(e)}]\{\Phi^{(e)}\}$ then

$$\int_{S_3^{(e)}} [N^{(e)}]^\mathrm{T} h(\phi - \phi_\infty) \mathrm{d}S$$

$$= \int_{S_3^{(e)}} h[N^{(e)}]^\mathrm{T}[N^{(e)}]\{\Phi^{(e)}\} \mathrm{d}S - \int_{S_3^{(e)}} h\phi_\infty [N^{(e)}]^\mathrm{T} \mathrm{d}S \qquad [6.40]$$

As a result, Eqs 6.35, 6.39 and 6.40 may be recombined to give the general element equations for field problems:

$$- \{G^{(e)}\} = \left(\int_{V^{(e)}} [B^{(e)}]^\mathrm{T}[D^{(e)}][B^{(e)}] \mathrm{d}V \right.$$

$$+ \int_{S_3^{(e)}} h[N^{(e)}]^\mathrm{T}[N^{(e)}] \mathrm{d}S \bigg) \{\Phi^{(e)}\} - \int_{V^{(e)}} Q^{(e)}[N^{(e)}]^\mathrm{T} \mathrm{d}V$$

$$+ \int_{S_2^{(e)}} q^{(e)}[N^{(e)}]^\mathrm{T} \mathrm{d}S - \int_{S_3^{(e)}} h\phi_\infty [N^{(e)}]^\mathrm{T} \mathrm{d}S \qquad [6.41]$$

Equation 6.41 consists of two parts, the elemental stiffness matrix $[k^{(e)}]$ and the force vector $\{F^{(e)}\}$, that is

$$- \{G^{(e)}\} = [k^{(e)}]\{\Phi^{(e)}\} - \{F^{(e)}\} \qquad [6.42]$$

and from Eq. 6.27,

$$\sum_{e=1}^{E} \{G^{(e)}\} = \sum_{e=1}^{E} ([k^{(e)}]\{\Phi^{(e)}\} - \{F^{(e)}\}) = 0 \qquad [6.43]$$

6.2.3 Summary

The two approaches give exactly the same final element equations, as would be expected (Eqs 6.17 and 6.41). The variational formulation appears to be the simpler technique, once the functional has been derived. However, the weighted residual method does not need a functional; it starts directly from the governing equations of the problem, and is therefore the more general approach.

In summary, the application of the finite element method to field problems requires that

$$\sum_{e=1}^{E} ([k^{(e)}]\{\Phi^{(e)}\} - \{F^{(e)}\}) = 0 \qquad [6.44]$$

where

$$[k^{(e)}] = \int_{V^{(e)}} [B^{(e)}]^{\mathrm{T}}[D^{(e)}][B^{(e)}]\mathrm{d}V + \int_{S_3^{(e)}} h[N^{(e)}]^{\mathrm{T}}[N^{(e)}]\mathrm{d}S \quad [6.45]$$

$$\{F^{(e)}\} = \int_{V^{(e)}} Q^{(e)}[N^{(e)}]^{\mathrm{T}}\mathrm{d}V - \int_{S_2^{(e)}} q^{(e)}[N^{(e)}]^{\mathrm{T}}\mathrm{d}S$$

$$+ \int_{S_3^{(e)}} h\phi_\infty [N^{(e)}]^{\mathrm{T}}\mathrm{d}S \quad [6.46]$$

Equation 6.44 can be used for one-, two- and three-dimensional problems, but needs to be tailored to each particular element type. Thermal problems are examined in detail in the next section, and the element equations for one-, two- and three-dimensional and axisymmetric elements are derived. For simplicity, the superscript (e) is dropped where only one element is being considered.

6.3   Thermal problems

The characteristic element equation Eq. 6.44 is directly applicable to heat transfer problems. The equation shows that the stiffness matrix can comprise two parts, one from the internal conduction effects in the element (through the thermal conductivities K_x, K_y and K_z represented in $[D]$) and the other from any convective heat transfer occurring on any surface of the element (convection coefficient h). The convection also gives rise to a term in the force vector. Note that every surface of the element that experiences convection will contribute one term to the stiffness matrix and one term to the force vector.

The other thermal effects occurring at each element, namely the internal heat source or sink Q and the applied surface heat flux q, each produce a single term in the force vector.

6.3.1   One-dimensional heat transfer

For the one-dimensional case, the differential equation simply reduces to

$$K \frac{\mathrm{d}^2\phi}{\mathrm{d}x^2} + Q = 0$$

with possible boundary conditions of

$$\phi = \phi_0$$

$$K \frac{\mathrm{d}\phi}{\mathrm{d}x} l_x + h(\phi - \phi_\infty) + q = 0$$

on the free surfaces.

A common example of a one-dimensional heat transfer problem is a fin, one end of which is connected to a heat source whose temperature is

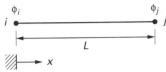

Figure 6.1 Linear one-dimensional element

known. Heat is lost to the surroundings through the free end and the perimeter surface of the fin.

Consider the development of the equations of a general one-dimensional element (with a linear interpolation function) for use in the thermal analysis of problems as described above, and illustrated in Fig. 6.1. The element was introduced in Chapter 4, and the interpolation function was shown to be (Eq. 4.4)

$$\phi = N_i\phi_i + N_j\phi_j = [N]\{\Phi\}$$

where $N_i = (1 - x/L)$ and $N_j = x/L$.

To derive the element equations for this particular element, its properties are substituted into the characteristic equation Eq. 6.44. The $[D]$ matrix is replaced with the constant K and the $[B]$ matrix is calculated by the correct differentiation of the shape functions in $[N]$. Remember that

$$\{g\} = \frac{d\phi}{dx} = -\frac{1}{L}\phi_i + \frac{1}{L}\phi_j = \left[-\frac{1}{L} \quad \frac{1}{L}\right]\left\{\begin{matrix}\phi_i\\\phi_j\end{matrix}\right\} = [B]\left\{\begin{matrix}\phi_i\\\phi_j\end{matrix}\right\}$$

Hence

$$[B] = \frac{1}{L}[-1 \ 1]\tag{6.47}$$

Therefore

$$\int_V [B]^T[D][B]dV = \int_0^L \frac{1}{L}\left\{\begin{matrix}-1\\1\end{matrix}\right\} K \frac{1}{L}[-1 \ 1]A \ dx$$

$$= \frac{KA}{L}\begin{bmatrix}1 & -1\\-1 & 1\end{bmatrix}\tag{6.48}$$

(This assumes the element has a constant cross-sectional area A; the effect of a variable cross-section is discussed after Example 6.1.)

For convection occurring around the perimeter of the element, a stiffness matrix term and a force vector term need to be calculated. For the first of these,

$$\int_S h[N]^T[N]dS = \int_0^L h \begin{bmatrix}N_i^2 & N_iN_j\\N_jN_i & N_j^2\end{bmatrix} P \ dx$$

where $dS = Pdx$ is substituted for the surface area, and P is the perimeter length which does not vary down the length of the element. The integral is easily evaluated using the factorial integration formula Eq. 4.20, and is found to be

$$\frac{hPL}{6}\begin{bmatrix}2 & 1\\1 & 2\end{bmatrix}\tag{6.49}$$

The force vector term from the convection around the perimeter is

$$\int_S h\phi_\infty [N]^T dS = \int_0^L h\phi_\infty \begin{Bmatrix} N_i \\ N_j \end{Bmatrix} P \, dx$$

again by substituting $dS = P \, dx$. Using local coordinates, this is worked out as

$$\frac{h\phi_\infty PL}{2} \begin{Bmatrix} 1 \\ 1 \end{Bmatrix} \tag{6.50}$$

For the convection that occurs at the end of the element (considering the end at node j), the stiffness matrix term is found from

$$\int_S h[N]^T [N] ds = \int_A h \begin{bmatrix} N_i^2 & N_i N_j \\ N_j N_i & N_j^2 \end{bmatrix} dA$$

where A is the cross-sectional area of the element. Because $N_i = 0$ and $N_j = 1$ at node j, the integral becomes

$$hA \begin{bmatrix} 0 & 0 \\ 0 & 1 \end{bmatrix} \tag{6.51}$$

The term in the force vector from the end convection is found from

$$\int_A h\phi_\infty [N]^T dA = h\phi_\infty A \begin{Bmatrix} 0 \\ 1 \end{Bmatrix} \tag{6.52}$$

The heat flux q must also be considered to act on either the perimeter or the end of the element. Each situation produces a term in the force vector, and is calculated in a similar manner to the corresponding term produced by the convection effect, but with $h\phi_\infty$ replaced by $-q$.

Therefore, for heat flux on the node j end of the element include (from Eq. 6.52)

$$-qA \begin{Bmatrix} 0 \\ 1 \end{Bmatrix} \tag{6.53}$$

in the force vector, and for heat flux around the perimeter include (from Eq. 6.50)

$$-\frac{qPL}{2} \begin{Bmatrix} 1 \\ 1 \end{Bmatrix} \tag{6.54}$$

Finally, the internal heat source Q will produce a similar term in the force vector calculated by

$$\int_V Q[N]^T dV = \int_0^L Q \begin{Bmatrix} N_i \\ N_j \end{Bmatrix} A \, dx$$

which with the use of the local integration formula is easily found to be

$$\frac{QAL}{2}\begin{Bmatrix} 1 \\ 1 \end{Bmatrix} \tag{6.55}$$

So the heat generated within the element is allocated equally to the two nodes.

In summary, for a general one-dimensional element, from Eqs 6.48, 6.49 and 6.51 the stiffness matrix may consist of

$$\left(\underbrace{\frac{KA}{L}\begin{bmatrix} 1 & -1 \\ -1 & 1 \end{bmatrix}}_{\substack{\text{from axial} \\ \text{conduction}}} + \underbrace{\frac{hPL}{6}\begin{bmatrix} 2 & 1 \\ 1 & 2 \end{bmatrix}}_{\substack{\text{perimeter} \\ \text{convection}}} + \underbrace{hA\begin{bmatrix} 0 & 0 \\ 0 & 1 \end{bmatrix}}_{\substack{\text{end} \\ \text{convection at} \\ \text{node } j}} \right) \tag{6.56}$$

An equivalent term to the third one would also appear if convection occurred at node i.

The force vector may consist of

$$\left(\underbrace{\frac{h\phi_\infty PL}{2}\begin{Bmatrix} 1 \\ 1 \end{Bmatrix}}_{\substack{\text{perimeter} \\ \text{convection}}} + \underbrace{h\phi_\infty A\begin{Bmatrix} 0 \\ 1 \end{Bmatrix}}_{\substack{\text{end} \\ \text{convection} \\ \text{at node } j}} - \underbrace{\frac{qPL}{2}\begin{Bmatrix} 1 \\ 1 \end{Bmatrix}}_{\substack{\text{perimeter} \\ \text{heat flux}}} - \underbrace{qA\begin{Bmatrix} 0 \\ 1 \end{Bmatrix}}_{\substack{\text{heat flux} \\ \text{at node } j}} \right.$$

$$\left. + \underbrace{\frac{QAL}{2}\begin{Bmatrix} 1 \\ 1 \end{Bmatrix}}_{\substack{\text{any internal} \\ \text{heat source}}} \right) \tag{6.57}$$

and again, convection or heat flux at node i would add further terms.

Example 6.1: straight fin analysis

Find the temperature distribution in the one-dimensional fin shown in Fig. 6.2.

The element matrices for the first three elements are identical. For example, for element (1), the stiffness matrix comprises the first two terms of Eq. 6.56, giving

$$[k^{(1)}] = \frac{AK}{L}\begin{bmatrix} 1 & -1 \\ -1 & 1 \end{bmatrix} + \frac{hPL}{6}\begin{bmatrix} 2 & 1 \\ 1 & 2 \end{bmatrix} \tag{6.58}$$

$$= 28\begin{bmatrix} 1 & -1 \\ -1 & 1 \end{bmatrix} + 16.67\begin{bmatrix} 2 & 1 \\ 1 & 2 \end{bmatrix} = \begin{bmatrix} 61.34 & -11.33 \\ -11.33 & 61.34 \end{bmatrix}$$

and the force vector consists of the first term of Eq. 6.57, namely

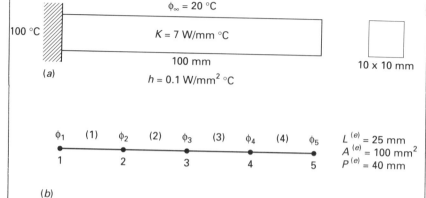

Figure 6.2 (a) One-dimensional fin (b) finite element idealization

$$\{F^{(1)}\} = \frac{h\phi_\infty PL}{2}\begin{Bmatrix} 1 \\ 1 \end{Bmatrix} = 1000 \begin{Bmatrix} 1 \\ 1 \end{Bmatrix} \qquad [6.59]$$

The fourth element experiences the same conditions, but in addition may lose heat through the right end at node j. Its stiffness matrix must therefore include an extra term (from Eq. 6.56), namely

$$hA \begin{bmatrix} 0 & 0 \\ 0 & 1 \end{bmatrix} \qquad [6.60]$$

so that

$$[k^{(4)}] = \begin{bmatrix} 61.34 & -11.33 \\ -11.33 & 61.34 \end{bmatrix} + 10 \begin{bmatrix} 0 & 0 \\ 0 & 1 \end{bmatrix}$$

$$= \begin{bmatrix} 61.34 & -11.33 \\ -11.33 & 71.34 \end{bmatrix}$$

Similarly, its force vector needs the extra term

$$h\phi_\infty A \begin{Bmatrix} 0 \\ 1 \end{Bmatrix} = 200 \begin{Bmatrix} 0 \\ 1 \end{Bmatrix} \qquad [6.61]$$

and therefore

$$\{F^{(4)}\} = 1000 \begin{Bmatrix} 1 \\ 1 \end{Bmatrix} + 200 \begin{Bmatrix} 0 \\ 1 \end{Bmatrix} = \begin{Bmatrix} 1000 \\ 1200 \end{Bmatrix}$$

The element matrices are now assembled to give the system equations, the boundary conditions are inserted, and the resulting set of simultaneous equations is then solved to give the unknown nodal temperatures. These steps are discussed in detail in Chapter 7.

The final global matrix is

$$
\begin{bmatrix}
61.34 & -11.33 & 0 & 0 & 0 \\
-11.33 & 122.68 & -11.33 & 0 & 0 \\
0 & -11.33 & 122.68 & -11.33 & 0 \\
0 & 0 & -11.33 & 122.68 & -11.33 \\
0 & 0 & 0 & -11.33 & 71.34
\end{bmatrix}
\begin{Bmatrix}
\phi_1 \\ \phi_2 \\ \phi_3 \\ \phi_4 \\ \phi_5
\end{Bmatrix}
$$

$$
= \begin{Bmatrix}
1000 \\ 2000 \\ 2000 \\ 2000 \\ 1200
\end{Bmatrix}
$$

However, after modification for the boundary condition $\phi_1 = 100\,°\text{C}$, the equations become

$$
\begin{bmatrix}
61.34 & -11.33 & 0 & 0 & 0 \\
0 & 122.68 & -11.33 & 0 & 0 \\
0 & -11.33 & 122.68 & -11.33 & 0 \\
0 & 0 & -11.33 & 122.68 & -11.33 \\
0 & 0 & 0 & -11.33 & 71.34
\end{bmatrix}
\begin{Bmatrix}
100 \\ \phi_2 \\ \phi_3 \\ \phi_4 \\ \phi_5
\end{Bmatrix}
$$

$$
= \begin{Bmatrix}
1000 + Q_1 \\ 3133 \\ 2000 \\ 2000 \\ 1200
\end{Bmatrix}
$$

where Q_1 is the heat applied from the surroundings at node 1, to maintain it at a constant temperature of 100 °C. Solution gives

$$\{\Phi_S\}^{\mathrm{T}} = [100.0\ \ 27.45\ \ 20.69\ \ 20.06\ \ 20.01]$$

and the heat flow at the left end is then calculated from the first equation to be $Q_1 = 4823$ watts.

A comparison of the results with those calculated theoretically is as follows:

Temperatures (°C)	ϕ_1	ϕ_2	ϕ_3	ϕ_4	ϕ_5
FE results	100.00	27.45	20.69	20.06	20.01
Theory	100.00	32.08	21.83	20.28	20.07
x location (mm)	0	25.0	50.0	75.0	100.0

The results are most inaccurate near the left-hand end of the fin, where the change in temperature gradient is the greatest, and hence a linear approximation is least accurate. A remeshing of the problem with smaller elements nearer that end would improve the results. However, in practice either more than four elements would be used to model the fin, or more sophisticated elements with higher-order interpolation functions would be chosen. (This problem is re-examined in Chapter 8 where it is modelled with quadratic elements, giving a substantial increase in the accuracy.)

Tapered fin analysis

In the derivation of the element equations it was assumed that the area was constant down the length of each element. If this is not the case the evaluation of the terms involving the area is not straightforward; the area needs to be expressed as a function of the x coordinate, and then integrated accordingly.

The simplest case is where the area varies linearly from one end of the element to the other. This is the same form of variation as that assumed for the temperature distribution in the x direction, and consequently the same shape functions can be applied. For example, if the area at node i is A_i and that at node j is A_j, then the area at a distance x from node i is

$$A = N_i A_i + N_j A_j = [N] \left\{ \begin{array}{c} A_i \\ A_j \end{array} \right\} \qquad [6.62]$$

Owing to the change in area, the stiffness matrix must be recalculated. The first term involves

$$\int_V [B]^T [D][B] dV = \frac{K}{L^2} \left[\begin{array}{cc} 1 & -1 \\ -1 & 1 \end{array} \right] \int_0^L A \, dx$$

Using Eq. 6.62 and the local integration formula of Eq. 4.20,

$$\int_0^L A dx = \int_0^L [N] \left\{ \begin{array}{c} A_i \\ A_j \end{array} \right\} dx = \frac{L}{2} [1 \quad 1] \left\{ \begin{array}{c} A_i \\ A_j \end{array} \right\}$$

Therefore the term becomes

$$\frac{K}{L} \frac{(A_i + A_j)}{2} \left[\begin{array}{cc} 1 & -1 \\ -1 & 1 \end{array} \right] = \frac{K\bar{A}}{L} \left[\begin{array}{cc} 1 & -1 \\ -1 & 1 \end{array} \right] \qquad [6.63]$$

where $\bar{A}$ is the average area. The matrix is identical to Eq. 6.48 except that the area used is now the average value.

With varying cross-sectional area, the element's perimeter will also vary. Again if the change is linear, we can write

$$P = N_i P_i + N_j P_j \qquad [6.64]$$

where P_i and P_j are the perimeters at the two nodes.

The convection from the perimeter adds a further term to the stiffness matrix. As in the derivation of Eq. 6.49, with $dS = P\,dx$,

$$\int_S h[N]^T[N]dS = \int_0^L h[N]^T[N]P\,dx$$

or

$$\int_0^L h \begin{bmatrix} N_i^2 P & N_i N_j P \\ N_j N_i P & N_j^2 P \end{bmatrix} dx$$

$$= \int_0^L h \begin{bmatrix} N_i^3 P_i + N_i^2 N_j P_j & N_i^2 N_j P_i + N_i N_j^2 P_j \\ N_i^2 N_j P_i + N_i N_j^2 P_j & N_i N_j^2 P_i + N_j^3 P_j \end{bmatrix} dx$$

Through the use of the factorial integration formula Eq. 4.20, this is evaluated to be

$$\frac{hL}{12} \begin{bmatrix} 3P_i + P_j & P_i + P_j \\ P_i + P_j & P_i + 3P_j \end{bmatrix} \tag{6.65}$$

The force vector term due to the convection around the perimeter also needs a similar approach, using $dS = P\,dx$:

$$\int_S h\phi_\infty[N]^T dS = \int_0^L h\phi_\infty \begin{Bmatrix} N_i P \\ N_j P \end{Bmatrix} dx = \frac{h\phi_\infty L}{6} \begin{Bmatrix} 2P_i + P_j \\ P_i + 2P_j \end{Bmatrix} \tag{6.66}$$

Therefore it is clear that it is not just a question of substituting the average perimeter length into Eqs 6.56 and 6.57 to obtain the element equations for a tapered element.

The other terms in Eqs 6.56 and 6.57 are dealt with as necessary. Where convection or heat flux occurs at one end of the element, the appropriate area is inserted rather than the common area previously used. The analysis of a tapered fin then proceeds in the same manner as for a straight one.

Example 6.2: tapered fin analysis

Calculate the temperature distribution in the tapered fin shown in Fig. 6.3.

Since the fin is tapered, each of the four element equations needed to describe the behaviour of the fin will be different, unlike the straight version of the fin described in Example 6.1.

For elements (1), (2) and (3), using Eqs 6.63 and 6.65,

$$[k] = \frac{K}{L} \frac{(A_i + A_j)}{2} \begin{bmatrix} 1 & -1 \\ -1 & 1 \end{bmatrix} + \frac{hL}{12} \begin{bmatrix} 3P_i + P_j & P_i + P_j \\ P_i + P_j & P_i + 3P_j \end{bmatrix} \tag{6.67}$$

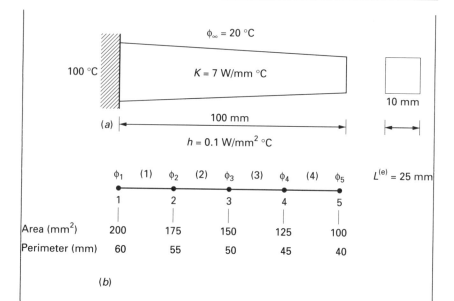

Figure 6.3

(b)

Therefore, for element (1),

$$[k^{(1)}] = 52.5 \begin{bmatrix} 1 & -1 \\ -1 & 1 \end{bmatrix} + 0.208 \begin{bmatrix} 235 & 115 \\ 115 & 225 \end{bmatrix}$$

$$= \begin{bmatrix} 101.38 & -28.58 \\ -28.58 & 99.30 \end{bmatrix}$$

The force vector for the first three elements is defined by Eq. 6.66 as

$$\{F\} = \frac{h\phi_\infty L}{6} \begin{Bmatrix} 2P_i + P_j \\ P_i + 2P_j \end{Bmatrix} \tag{6.68}$$

For element (1) this becomes

$$\{F^{(1)}\} = 8.333 \begin{Bmatrix} 175 \\ 170 \end{Bmatrix}$$

The element equations for elements (2) and (3) are calculated in the same way from Eqs 6.67 and 6.68. However, element (4) has an extra term in its stiffness matrix and force vector due to the convection experienced at its free end. In the stiffness matrix the additional term is

$$hA^{(4)} \begin{bmatrix} 0 & 0 \\ 0 & 1 \end{bmatrix} = 10 \begin{bmatrix} 0 & 0 \\ 0 & 1 \end{bmatrix}$$

and in the force vector the additional term is

$$h\phi_\infty A^{(4)} \begin{Bmatrix} 0 \\ 1 \end{Bmatrix} = 200 \begin{Bmatrix} 0 \\ 1 \end{Bmatrix}$$

When calculated, the complete set of element equations is found to be as follows:

$$[k^{(1)}] = \begin{bmatrix} 101.38 & -28.58 \\ -28.58 & 99.30 \end{bmatrix}$$

$$[k^{(2)}] = \begin{bmatrix} 90.22 & -23.66 \\ -23.66 & 88.14 \end{bmatrix}$$

$$[k^{(3)}] = \begin{bmatrix} 78.96 & -18.64 \\ -18.64 & 76.88 \end{bmatrix}$$

$$[k^{(4)}] = \begin{bmatrix} 67.90 & -13.82 \\ -13.82 & 75.82 \end{bmatrix}$$

$$\{F^{(1)}\} = 8.33 \begin{Bmatrix} 175 \\ 170 \end{Bmatrix}$$

$$\{F^{(2)}\} = 8.33 \begin{Bmatrix} 160 \\ 155 \end{Bmatrix}$$

$$\{F^{(3)}\} = 8.33 \begin{Bmatrix} 145 \\ 140 \end{Bmatrix}$$

$$\{F^{(4)}\} = \begin{Bmatrix} 1083.29 \\ 1241.65 \end{Bmatrix}$$

These combine to give the following system equations:

$$\begin{bmatrix} 101.38 & -28.58 & 0 & 0 & 0 \\ -28.58 & 189.52 & -23.66 & 0 & 0 \\ 0 & -23.66 & 167.10 & -18.64 & 0 \\ 0 & 0 & -18.64 & 144.78 & -13.82 \\ 0 & 0 & 0 & -13.82 & 75.82 \end{bmatrix} \begin{Bmatrix} \phi_1 \\ \phi_2 \\ \phi_3 \\ \phi_4 \\ \phi_5 \end{Bmatrix}$$

$$= \begin{Bmatrix} 1458.28 \\ 2749.89 \\ 2499.90 \\ 2249.91 \\ 1241.65 \end{Bmatrix}$$

When these equations are modified to take account of the boundary condition $\phi_1 = 100\,°C$ and are then solved, the following distribution is predicted:

$$\{\Phi_s\}^{T} = [100.00\ 32.31\ 21.80\ 20.26\ 20.07]\ °C$$

Using these values, the heat flow applied at node 1 to maintain the temperature at 100 °C is calculated to be 7737 watts. It is interesting to compare this with the value of 4823 watts produced by the straight fin analysis in Example 6.1. Clearly the tapered fin is more efficient at dissipating heat than the straight version.

Calculation of internal heat flows

By Fourier's law, the quantity of heat q crossing a unit area per unit time in the x direction is given by

$$q = - K \frac{d\phi}{dx} \qquad [6.69]$$

where K is the thermal conductivity. Therefore, once the temperature distribution is calculated within a body, it is possible to calculate the heat flow through each element using Eq. 6.69 multiplied by the cross-sectional area.

In one dimension, the quantity $d\phi/dx$ is defined as the gradient vector $\{g\}$ by Eq. 6.9 and is shown in Eq. 6.15 to be

$$\{g^{(e)}\} = [B^{(e)}]\{\Phi^{(e)}\}$$

Hence the heat flux through the element can be found from

$$q_{\text{flux}}^{(e)} = - K[B^{(e)}]\{\Phi^{(e)}\} \qquad [6.70]$$

Since the element examined in this section is a simplex element, *i.e.* it uses a linear variation of temperature, a constant heat flux is predicted through the element (because $d\phi/dx$ is constant). With higher-order elements, where the interpolation function might be quadratic or cubic, the predicted heat flow would be linear or quadratic respectively.

Example 6.3: heat flow calculations

For the fin analysed in Example 6.1, calculate the heat flow rates in each element.

The heat flux is found from

$$q_{\text{flux}}^{(e)} = - K[B^{(e)}]\{\Phi^{(e)}\}$$

For a one-dimensional element with a linear interpolation function,

$$[B] = \frac{1}{L}[-1 \ 1]$$

The elemental lengths are all equal, and so the $[B]$ matrix is the same for the elements. Considering element (1),

$$[B^{(1)}] = 0.4 \, [-1 \ 1]$$

Therefore

$$q_{\text{flux}}^{(1)} = -7 \times 0.04 \, [-1 \ 1] \left\{ \begin{array}{c} 100 \\ 27.45 \end{array} \right\} = 20.31 \text{ watts/mm}^2$$

The area of each element is 100 mm^2, so the heat flow in element (1) is

$$Q_{\text{flow}}^{(1)} = 2031.4 \text{ watts}$$

Similarly,

$$Q_{\text{flow}}^{(2)} = 189.28 \text{ watts}$$
$$Q_{\text{flow}}^{(3)} = 17.64 \text{ watts}$$
$$Q_{\text{flow}}^{(4)} = 1.40 \text{ watts}$$

Composite wall analysis

The one-dimensional element equations derived in this chapter can also be used very effectively to study the thermal distribution through composite (layered) walls, where heat flow only occurs in one direction. The following example illustrates the principles well.

Example 6.4: composite wall analysis
Determine the temperature distributions in the wall shown in Fig. 6.4, and calculate the heat flow through the wall thickness.

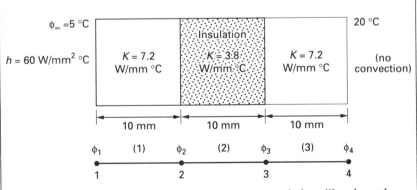

Figure 6.4

Element (1) experiences convection at one end, but like the other elements no perimeter convection. The stiffness matrix for element

(1) is therefore found (from Eq. 6.56) to be

$$[k^{(1)}] = \frac{KA}{L} \begin{bmatrix} 1 & -1 \\ -1 & 1 \end{bmatrix} + hA \begin{bmatrix} 1 & 0 \\ 0 & 0 \end{bmatrix} \qquad [6.71]$$

$$= 0.72 \begin{bmatrix} 1 & -1 \\ -1 & 1 \end{bmatrix} + 60 \begin{bmatrix} 1 & 0 \\ 0 & 0 \end{bmatrix} = \begin{bmatrix} 60.72 & -0.72 \\ -0.72 & 0.72 \end{bmatrix}$$

$$[6.72]$$

where a unit cross-sectional area is assumed for the element. The force vector term comprises just one term, for the end convection:

$$\{F^{(1)}\} = h\phi_\infty A \begin{Bmatrix} 1 \\ 0 \end{Bmatrix} = 60 \times 5 \begin{Bmatrix} 1 \\ 0 \end{Bmatrix} \qquad [6.73]$$

For elements (2) and (3), the stiffness matrix consists of only one term, and the force vector term is zero (no terms in Eq. 6.57 apply). Specifically,

$$[k^{(2)}] = 0.038 \begin{bmatrix} 1 & -1 \\ -1 & 1 \end{bmatrix} \qquad \{F^{(2)}\} = \begin{Bmatrix} 0 \\ 0 \end{Bmatrix} \qquad [6.74]$$

$$[k^{(3)}] = 0.072 \begin{bmatrix} 1 & -1 \\ -1 & 1 \end{bmatrix} \qquad \{F^{(3)}\} = \begin{Bmatrix} 0 \\ 0 \end{Bmatrix} \qquad [6.75]$$

Assembly of Eqs 6.72 to 6.75 gives

$$\begin{bmatrix} 60.72 & -0.72 & 0.0 & 0.0 \\ -0.72 & 0.758 & -0.038 & 0.0 \\ 0.0 & -0.038 & 0.758 & -0.72 \\ 0.0 & 0.0 & -0.72 & 0.72 \end{bmatrix} \begin{Bmatrix} \phi_1 \\ \phi_2 \\ \phi_3 \\ \phi_4 \end{Bmatrix} = \begin{Bmatrix} 300 \\ 0 \\ 0 \\ 0 \end{Bmatrix}$$

Application of the boundary condition $\phi_4 = 20\ °C$ yields

$$\begin{bmatrix} 60.72 & -0.72 & 0.0 & 0.0 \\ -0.72 & 0.758 & -0.038 & 0.0 \\ 0.0 & -0.038 & 0.758 & -0.72 \\ 0.0 & 0.0 & -0.72 & 0.72 \end{bmatrix} \begin{Bmatrix} \phi_1 \\ \phi_2 \\ \phi_3 \\ 20 \end{Bmatrix} = \begin{Bmatrix} 300 \\ 0 \\ 0 \\ Q_4 \end{Bmatrix}$$

where Q_4 is the heat applied at node 4 to maintain the temperature at 20 °C.

Solution of the equations gives

$$\{\Phi_S\}^T = [5.008 \ 5.724 \ 19.284 \ 20.0] \ °C$$

and $Q_4 = 0.516$ watts.

One would expect the heat flow through each element to be the same, and indeed Eq. 6.70 can be used to prove this is the case. The calculated values for temperature and heat flow compare exactly with those derived theoretically. The reason is that the temperature distribution through such a wall is linear, and hence the finite element analysis, with linear interpolation elements, models the situation exactly. Analysing the wall with more elements would give no further information.

6.3.2 Two-dimensional heat transfer

The basic element in two dimensions is the triangle, as introduced in Chapter 4 and illustrated in Fig. 6.5. The distribution of temperature (assuming a linear variation) over the element is (Eq. 4.9)

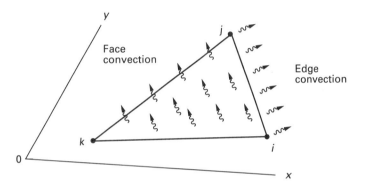

Figure 6.5 Linear two-dimensional thermal element, showing edge and face convection

$$\phi = N_i\phi_i + N_j\phi_j + N_k\phi_k = [N]\{\Phi\} \tag{6.76}$$

where

$$\begin{Bmatrix} N_i \\ N_j \\ N_k \end{Bmatrix} = \frac{1}{2A} \begin{bmatrix} a_i & b_i & c_i \\ a_j & b_j & c_j \\ a_k & b_k & c_k \end{bmatrix} \begin{Bmatrix} 1 \\ x \\ y \end{Bmatrix} \tag{6.77}$$

and the a, b and c are constants calculated from the coordinates of the element as derived in Section 4.3.

As for the one-dimensional case, the element equations are found by deriving the necessary matrices in Eq. 6.44 and making the substitutions. Firstly,

$$\{g\} = \begin{Bmatrix} \dfrac{\partial\phi}{\partial x} \\ \dfrac{\partial\phi}{\partial y} \end{Bmatrix} = \frac{1}{2A} \begin{bmatrix} b_i & b_j & b_k \\ c_i & c_j & c_k \end{bmatrix} \begin{Bmatrix} \phi_i \\ \phi_j \\ \phi_k \end{Bmatrix} = [B] \begin{Bmatrix} \phi_i \\ \phi_j \\ \phi_k \end{Bmatrix} \tag{6.78}$$

Therefore

$$[B] = \frac{1}{2A} \begin{bmatrix} b_i & b_j & b_k \\ c_i & c_j & c_k \end{bmatrix} \begin{Bmatrix} \phi_i \\ \phi_j \\ \phi_k \end{Bmatrix} \qquad [6.79]$$

The material property matrix $[D]$ is defined as

$$[D] = \begin{bmatrix} K_x & 0 \\ 0 & K_y \end{bmatrix} \qquad [6.80]$$

The first term of the stiffness matrix is then found from

$$\int_V [B]^T [D][B] \mathrm{d}V$$

$$= \int_V \frac{1}{4A^2} \begin{bmatrix} b_i & c_i \\ b_j & c_j \\ b_k & c_k \end{bmatrix} \begin{bmatrix} K_x & 0 \\ 0 & K_y \end{bmatrix} \begin{bmatrix} b_i & b_j & b_k \\ c_i & c_j & c_k \end{bmatrix} t \, \mathrm{d}A$$

assuming that the element has constant thickness t. The integral contains constants and is readily evaluated to give

$$[B]^T [D][B] t A \qquad [6.81]$$

For an isotropic material with $K_x = K_y = K$, Eq. 6.81 reduces to

$$\frac{Kt}{4A} \begin{bmatrix} b_i^2 + c_i^2 & b_i b_j + c_i c_j & b_i b_k + c_i c_k \\ b_j b_i + c_j c_i & b_j^2 + c_j^2 & b_j b_k + c_j c_k \\ b_k b_i + c_k c_i & b_k b_j + c_k c_j & b_k^2 + c_k^2 \end{bmatrix} \qquad [6.82]$$

For a triangular element, convection is possible from its three edges and from its flat faces (Fig. 6.5), and each of these effects contributes one term to the stiffness matrix and one term to the force vector. Firstly, consider convection from edge ij of the element. The stiffness matrix term will be

$$\int_S h[N]^T [N] \mathrm{d}S = \int_S h \begin{bmatrix} N_i^2 & N_i N_j & N_i N_k \\ N_j N_i & N_j^2 & N_j N_k \\ N_k N_i & N_k N_j & N_k^2 \end{bmatrix} \mathrm{d}S$$

$$= \int_S h \begin{bmatrix} N_i^2 & N_i N_j & 0 \\ N_j N_i & N_j^2 & 0 \\ 0 & 0 & 0 \end{bmatrix} \mathrm{d}S$$

since $N_k = 0$ along side ij. The integral is over the edge of the element between nodes i and j, which for constant thickness t means area $H_{ij}t$, where H_{ij} is the distance between nodes i and j. Using the local integration formula Eq. 4.25 to evaluate this, as described in Section 4.4, gives

$$\int_{H_{ij}} h \begin{bmatrix} N_i^2 & N_iN_j & 0 \\ N_jN_i & N_j^2 & 0 \\ 0 & 0 & 0 \end{bmatrix} t \, dH = \frac{hH_{ij}t}{6} \begin{bmatrix} 2 & 1 & 0 \\ 1 & 2 & 0 \\ 0 & 0 & 0 \end{bmatrix} \qquad [6.83]$$

The force vector term produced by the convection effect is calculated from

$$\int_S h\phi_\infty [N]^T ds = \int_{H_{ij}} h\phi_\infty \begin{Bmatrix} N_i \\ N_j \\ N_k \end{Bmatrix} t \, dH$$

Using natural coordinates again, this results in

$$\frac{h\phi_\infty H_{ij}t}{2} \begin{Bmatrix} 1 \\ 1 \\ 0 \end{Bmatrix} \qquad [6.84]$$

If convection takes place on another side of the element, similar equations to Eqs 6.83 and 6.84 are generated with obvious rearrangement of the non-zero terms.

Where convection occurs from a face of the element, the stiffness matrix term is

$$\int_A h \begin{bmatrix} N_i^2 & N_iN_j & N_iN_k \\ N_jN_i & N_j^2 & N_jN_k \\ N_kN_i & N_kN_j & N_k^2 \end{bmatrix} dA = \frac{hA}{12} \begin{bmatrix} 2 & 1 & 1 \\ 1 & 2 & 1 \\ 1 & 1 & 2 \end{bmatrix} \qquad [6.85]$$

by the use of local coordinates in the usual manner.

The force vector part is

$$\int_A h\phi_\infty \begin{Bmatrix} N_i \\ N_j \\ N_k \end{Bmatrix} dA = \frac{h\phi_\infty A}{3} \begin{Bmatrix} 1 \\ 1 \\ 1 \end{Bmatrix} \qquad [6.86]$$

The other terms in the force vector are equally straightforward to calculate. Where there is an internal heat source,

$$\int_V Q[N]^T dV = \frac{QtA}{3} \begin{Bmatrix} 1 \\ 1 \\ 1 \end{Bmatrix} \qquad [6.87]$$

showing that the heat generated within the body is allocated equally to the three nodes.

Any incident heat flux on the element can occur over the element's edges or faces as with the convection effect, and similar terms to those in Eqs 6.84 and 6.86 are derived. For the edge between nodes i and j, the force vector term is

$$\int_S q[N]^T dS = \frac{qH_{ij}t}{2}\begin{Bmatrix} 1 \\ 1 \\ 0 \end{Bmatrix}$$ [6.88]

and for the face effect,

$$\int_A q[N]^T dA = \frac{qA}{3}\begin{Bmatrix} 1 \\ 1 \\ 1 \end{Bmatrix}$$ [6.89]

The element equations for the triangular element can therefore now be found by summing the stiffness matrix terms, namely Eqs 6.82, 6.83 and 6.85, to give

$$\left([B]^T[D][B]tA + \underset{\substack{\text{convection} \\ \text{at edge } ij}}{\frac{hH_{ij}t}{6}\begin{bmatrix} 2 & 1 & 0 \\ 1 & 2 & 0 \\ 0 & 0 & 0 \end{bmatrix}} + \underset{\substack{\text{face area} \\ \text{convection}}}{\frac{hA}{12}\begin{bmatrix} 2 & 1 & 1 \\ 1 & 2 & 1 \\ 1 & 1 & 2 \end{bmatrix}} \right)$$ [6.90]

(under first term: "from conduction")

Here the convection effects are only represented for side ij; any convection on side jk or ki introduces similar terms.

The force vector (with convection and heat flux terms similarly only considered for side ij) consists of Eqs 6.84 and 6.86 to 6.89, producing

$$\left(\underset{\substack{\text{convection} \\ \text{at edge } ij}}{\frac{h\phi_\infty H_{ij}t}{2}\begin{Bmatrix} 1 \\ 1 \\ 0 \end{Bmatrix}} + \underset{\substack{\text{face area} \\ \text{convection}}}{\frac{h\phi_\infty A}{3}\begin{Bmatrix} 1 \\ 1 \\ 1 \end{Bmatrix}} + \underset{\substack{\text{internal} \\ \text{heat source}}}{\frac{QtA}{3}\begin{Bmatrix} 1 \\ 1 \\ 1 \end{Bmatrix}} - \underset{\substack{\text{heat flux} \\ \text{on edge } ij}}{\frac{qH_{ij}t}{2}\begin{Bmatrix} 1 \\ 1 \\ 0 \end{Bmatrix}} \right.$$

$$\left. - \underset{\substack{\text{heat flux} \\ \text{on face} \\ \text{area}}}{\frac{qA}{3}\begin{Bmatrix} 1 \\ 1 \\ 1 \end{Bmatrix}} \right)$$ [6.91]

Example 6.5: two-dimensional heat transfer

Calculate the element equations for the element shown in Fig. 6.6, which experiences convection on side jk and its upper face.

There will be three terms in the stiffness matrix, one from the normal conduction part and one from each of the convection effects. The first of these is calculated from Eq. 6.82, which needs the values

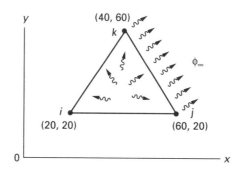

Figure 6.6

of the a, b and c constants. From Eq. 4.8,

$$b_i = y_j - y_k = -40 \qquad c_i = x_k - x_j = -20$$
$$b_j = y_k - y_i = 40 \qquad c_j = x_i - x_k = -20$$
$$b_k = y_i - y_j = 0 \qquad c_k = x_j - x_i = 40$$

and the area equals 800 mm^2.

Substituting into Eq. 6.82 gives

$$10^2 \times \frac{Kt}{4A} \begin{bmatrix} 20 & -12 & -8 \\ -12 & 20 & -8 \\ -8 & -8 & 16 \end{bmatrix} = \begin{bmatrix} 4.69 & -2.81 & -1.88 \\ -2.81 & 4.69 & -1.88 \\ -1.88 & -1.88 & 3.75 \end{bmatrix} \qquad [6.92]$$

For the convection on edge jk, the stiffness matrix will include a term (from Eq. 6.83)

$$\frac{hH_{jk}t}{6} \begin{bmatrix} 0 & 0 & 0 \\ 0 & 2 & 1 \\ 0 & 1 & 2 \end{bmatrix}$$

The length of the side between nodes j and k equals $\sqrt{2000}$; therefore

$$\frac{0.15 \times \sqrt{2000} \times 1}{6} \begin{bmatrix} 0 & 0 & 0 \\ 0 & 2 & 1 \\ 0 & 1 & 2 \end{bmatrix} = \begin{bmatrix} 0 & 0 & 0 \\ 0 & 2.24 & 1.12 \\ 0 & 1.12 & 2.24 \end{bmatrix} \qquad [6.93]$$

Equation 6.85 is the matrix that should be included when convection occurs from the face of an element. In this case,

$$\frac{hA}{12} \begin{bmatrix} 2 & 1 & 1 \\ 1 & 2 & 1 \\ 1 & 1 & 2 \end{bmatrix} = 10 \begin{bmatrix} 2 & 1 & 1 \\ 1 & 2 & 1 \\ 1 & 1 & 2 \end{bmatrix} \qquad [6.94]$$

The stiffness matrix for the two-dimensional element then equals the sum of Eqs 6.92 to 6.94, *i.e.*

$$\begin{bmatrix} 24.69 & 7.19 & 8.12 \\ 7.19 & 26.93 & 9.24 \\ 8.12 & 9.24 & 25.99 \end{bmatrix} \qquad\qquad [6.95]$$

The force vector for the element will consist of two terms. Firstly, for the edge convection, from Eq. 6.84

$$\frac{h\phi_\infty H_{jk}t}{2} \begin{Bmatrix} 0 \\ 1 \\ 1 \end{Bmatrix} = 67.08 \begin{Bmatrix} 0 \\ 1 \\ 1 \end{Bmatrix} \qquad\qquad [6.96]$$

Secondly, for the face convection, Eq. 6.86 gives

$$\frac{h\phi_\infty A}{3} \begin{Bmatrix} 1 \\ 1 \\ 1 \end{Bmatrix} = 800 \begin{Bmatrix} 1 \\ 1 \\ 1 \end{Bmatrix} \qquad\qquad [6.97]$$

Therefore, from Eqs 6.96 and 6.97, the force vector for the element is

$$\begin{Bmatrix} 800.00 \\ 867.08 \\ 867.08 \end{Bmatrix} \qquad\qquad [6.98]$$

So the element equations for the given element under the defined operating conditions are

$$\begin{bmatrix} 24.69 & 7.19 & 8.12 \\ 7.19 & 26.93 & 9.24 \\ 8.12 & 9.24 & 25.99 \end{bmatrix} \begin{Bmatrix} \phi_i \\ \phi_j \\ \phi_k \end{Bmatrix} = \begin{Bmatrix} 800.00 \\ 867.08 \\ 867.08 \end{Bmatrix}$$

6.3.3   Three-dimensional heat transfer

The derivation of the element equations for the simple three-dimensional element (the tetrahedron) with a linear interpolation function presents little further difficulty, particularly with the use of local volume coordinates to facilitate the integration. The basic element is shown in Fig. 6.7. The interpolation function was introduced in Eq. 4.14 as

$$\phi = N_i\phi_i + N_j\phi_j + N_k\phi_k + N_l\phi_l \qquad\qquad [6.99]$$

where the shape functions are defined by

$$\begin{Bmatrix} N_i \\ N_j \\ N_k \\ N_l \end{Bmatrix} = \frac{1}{6V} \begin{bmatrix} a_i & b_i & c_i & d_i \\ a_j & b_j & c_j & d_j \\ a_k & b_k & c_k & d_k \\ a_l & b_l & c_l & d_l \end{bmatrix} \begin{Bmatrix} 1 \\ x \\ y \\ z \end{Bmatrix} \qquad\qquad [6.100]$$

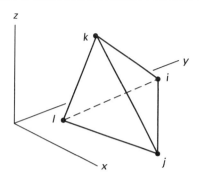

Figure 6.7 Linear three-dimensional thermal element

and the constants a, b, c and d are calculated directly from the nodal coordinates.

The $[B]$ matrix, found from the differentiation of Eq. 6.99 with 6.100, proves to be

$$[B] = \frac{1}{6V} \begin{bmatrix} b_i & b_j & b_k & b_l \\ c_i & c_j & c_k & c_l \\ d_i & d_j & d_k & d_l \end{bmatrix} \qquad [6.101]$$

Calculation of the stiffness matrix is then straightforward, and is found from the multiplication of the matrices

$$[k] = [B]^{\mathrm{T}}[D][B]V \qquad [6.102]$$

if the element has a volume V.

When there is convection over the side defined by nodes i, j and k, the stiffness matrix term is derived from

$$\int_S h\lfloor N\rfloor^{\mathrm{T}}\lfloor N\rfloor \mathrm{d}S = \int_S h \begin{bmatrix} N_i^2 & N_iN_j & N_iN_k & N_iN_l \\ N_jN_i & N_j^2 & N_jN_k & N_jN_l \\ N_kN_i & N_kN_j & N_k^2 & N_kN_l \\ N_lN_i & N_lN_j & N_lN_k & N_l^2 \end{bmatrix} \mathrm{d}S$$

However, on side ijk the shape function N_l corresponding to node l equals zero, and so the coefficients in the fourth row and column of the matrix must be set equal to zero. Therefore the integral becomes

$$\int_{A_{ijk}} h \begin{bmatrix} N_i^2 & N_iN_j & N_iN_k & 0 \\ N_jN_i & N_j^2 & N_jN_k & 0 \\ N_kN_i & N_kN_j & N_k^2 & 0 \\ 0 & 0 & 0 & 0 \end{bmatrix} \mathrm{d}A = \frac{hA_{ijk}}{12} \begin{bmatrix} 2 & 1 & 1 & 0 \\ 1 & 2 & 1 & 0 \\ 1 & 1 & 2 & 0 \\ 0 & 0 & 0 & 0 \end{bmatrix} \qquad [6.103]$$

The term produced in the force vector from this convection is

$$\int_S h\phi_\infty[N]^{\mathrm{T}}\mathrm{d}S = \int_{A_{ijk}} h\phi_\infty \begin{Bmatrix} N_i \\ N_j \\ N_k \\ 0 \end{Bmatrix} \mathrm{d}A = \frac{h\phi_\infty A_{ijk}}{3} \begin{Bmatrix} 1 \\ 1 \\ 1 \\ 0 \end{Bmatrix} \qquad [6.104]$$

where evaluation again is achieved easily by the use of local volume coordinates.

As usual, if convection occurs on other faces of the element, similar terms to those in Eqs 6.103 and 6.104 are produced, but with the obvious rearrangement of the non-zero coefficients.

The heat flux term is derived in the same way as Eq. 6.104. For example, if it acts over face ijk,

$$
\int_S q[N]^T dS = \int_{A_{ijk}} q \begin{Bmatrix} N_i \\ N_j \\ N_k \\ 0 \end{Bmatrix} dA = \frac{qA_{ijk}}{3} \begin{Bmatrix} 1 \\ 1 \\ 1 \\ 0 \end{Bmatrix}
$$

[6.105]

Finally, any internal heat source is included in the force vector by the term

$$
\int_V Q[N]^T dV = \int_V Q \begin{Bmatrix} N_i \\ N_j \\ N_k \\ N_l \end{Bmatrix} dV = \frac{QV}{4} \begin{Bmatrix} 1 \\ 1 \\ 1 \\ 1 \end{Bmatrix}
$$

[6.106]

In summary then, for an element with convection and incident heat flux acting over the face defined by nodes i, j and k, the stiffness matrix is

$$
\left([B]^T[D][B]V + \frac{hA_{ijk}}{12} \begin{bmatrix} 2 & 1 & 1 & 0 \\ 1 & 2 & 1 & 0 \\ 1 & 1 & 2 & 0 \\ 0 & 0 & 0 & 0 \end{bmatrix} \right)
$$

[6.107]

$$
\begin{matrix} \text{from} & \text{from convection} \\ \text{conduction} & \text{on face } ijk \end{matrix}
$$

and the force vector is

$$
\left(\frac{h\phi_\infty A_{ijk}}{3} \begin{Bmatrix} 1 \\ 1 \\ 1 \\ 0 \end{Bmatrix} - \frac{qA_{ijk}}{3} \begin{Bmatrix} 1 \\ 1 \\ 1 \\ 0 \end{Bmatrix} + \frac{QV}{4} \begin{Bmatrix} 1 \\ 1 \\ 1 \\ 1 \end{Bmatrix} \right)
$$

[6.108]

$$
\begin{matrix} \text{from} & \text{from heat} & \text{from any} \\ \text{convection on} & \text{flux on} & \text{internal} \\ \text{face } ijk & \text{face } ijk & \text{heat} \\ & & \text{source} \end{matrix}
$$

6.3.4   Axisymmetric heat transfer

If a three-dimensional body and the thermal distribution in that body are axisymmetric, then the situation may be modelled using axisymmetric elements. The basic linear element was first discussed in Section 4.6, and was shown to be an axisymmetric ring with a triangular cross-section defined in cylindrical coordinates, so that the symmetry is about the z axis, as shown in Fig. 6.8. The differential equation of heat conduction for such an axisymmetric case, in cylindrical coordinates, is

$$\frac{1}{r}\frac{\partial}{\partial r}\left(rK_r\frac{\partial\phi}{\partial r}\right) + \frac{\partial}{\partial z}\left(K_z\frac{\partial\phi}{\partial z}\right) + Q = 0 \qquad [6.109]$$

The boundary conditions are similar to those used for the general Cartesian system in Eqs 6.2 to 6.4, namely

$$\phi = \phi_1 \qquad [6.110]$$

on surface S_1, and

$$K_r\frac{\partial\phi}{\partial r}l_r + K_z\frac{\partial\phi}{\partial z}l_z + q = 0 \qquad [6.111]$$

$$K_r\frac{\partial\phi}{\partial r}l_r + K_z\frac{\partial\phi}{\partial z}l_z + h(\phi - \phi_\infty) = 0 \qquad [6.112]$$

on surfaces S_2 and S_3, where K_r and K_z are the thermal conductivities in the r and z directions respectively, l_r and l_z are the direction cosines of the outward normal to the surface, and the other terms have their usual meanings.

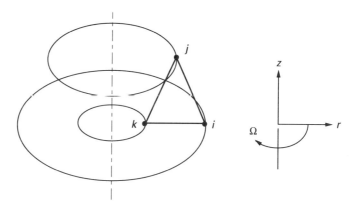

Figure 6.8 Axisymmetric thermal element

These equations are similar to those describing field problems in Cartesian coordinates (Eqs 6.1 to 6.4), except that the first term in Eq. 6.109 is a function of the radial position. The finite element equations for such problems are best derived by Galerkin's formulation in the same way as described in Section 6.2.2, but with some minor modifications.

The weighted residual method begins with the calculation of the residual integral of an element (e):

$$\{G^{(e)}\} = \int_{V^{(e)}} [N^{(e)}]^{\mathrm{T}} R^{(e)} \mathrm{d}V$$

$$= \int_{V^{(e)}} [N^{(e)}]^{\mathrm{T}} \left(\frac{K_r}{r} \frac{\partial}{\partial r} \left(r \frac{\partial \phi}{\partial r} \right) + K_z \frac{\partial^2 \phi}{\partial z^2} + Q \right) \mathrm{d}V \qquad [6.113]$$

The second derivative on the right-hand side of this equation must be replaced with first derivatives using integration by parts, as presented in Eq. 6.29, so that

$$[N^{(e)}]^{\mathrm{T}} K_z \frac{\partial^2 \phi}{\partial z^2} = K_z \frac{\partial}{\partial z} \left([N^{(e)}]^{\mathrm{T}} \frac{\partial \phi}{\partial z} \right) - K_z \frac{\partial [N^{(e)}]^{\mathrm{T}}}{\partial z} \frac{\partial \phi}{\partial z} \qquad [6.114]$$

The first term of the right-hand side of Eq. 6.113 can also be reformatted so that it is in a similar format to Eq. 6.114. This requires the following relationship:

$$[N^{(e)}]^{\mathrm{T}} \left(\frac{K_r}{r} \frac{\partial}{\partial r} \left(r \frac{\partial \phi}{\partial r} \right) \right)$$

$$= \frac{K_r}{r} \frac{\partial}{\partial r} \left([N^{(e)}]^{\mathrm{T}} r \frac{\partial \phi}{\partial r} \right) - K_r \frac{\partial [N^{(e)}]^{\mathrm{T}}}{\partial r} \frac{\partial \phi}{\partial r} \qquad [6.115]$$

If the last two equations are then substituted into Eq. 6.113, the following is obtained:

$$\{G^{(e)}\} = - \int_{V^{(e)}} \left(K_r \frac{\partial [N^{(e)}]^{\mathrm{T}}}{\partial r} \frac{\partial \phi}{\partial r} + K_z \frac{\partial [N^{(e)}]^{\mathrm{T}}}{\partial z} \frac{\partial \phi}{\partial z} \right) \mathrm{d}V$$

$$+ \int_{V^{(e)}} [N^{(e)}]^{\mathrm{T}} Q \, \mathrm{d}V + \int_{V^{(e)}} \left(\frac{K_r}{r} \frac{\partial}{\partial r} \left([N^{(e)}]^{\mathrm{T}} r \frac{\partial \phi}{\partial r} \right) \right.$$

$$+ K_z \frac{\partial}{\partial z} \left([N^{(e)}]^{\mathrm{T}} \frac{\partial \phi}{\partial z} \right) \Big) \mathrm{d}V \qquad [6.116]$$

The third integral in this equation can be replaced using Gauss' theorem, and yields

$$\int_{S^{(e)}} \left(\frac{K_r}{r} [N^{(e)}]^{\mathrm{T}} r \frac{\partial \phi}{\partial r} l_r + K_z [N^{(e)}]^{\mathrm{T}} \frac{\partial \phi}{\partial z} l_z \right) \mathrm{d}S \qquad [6.117]$$

where $S^{(e)}$ is a surface of the element, and l_r and l_z are the direction cosines of the normal to the surface. The variation of temperature can be expressed in the usual way in terms of the shape functions and nodal temperatures:

$$\phi = [N^{(e)}]\{\Phi^{(e)}\} \qquad [6.118]$$

Consequently, the weighted residual integral is finally calculated to be

$$\{G^{(e)}\} = - \left(\int_{V^{(e)}} \left(K_r \frac{\partial [N^{(e)}]^T}{\partial r} \frac{\partial [N^{(e)}]}{\partial r} \right. \right.$$

$$\left. + K_z \frac{\partial [N^{(e)}]^T}{\partial z} \frac{\partial [N^{(e)}]}{\partial z} \right) dV \right) \{\Phi^{(e)}\}$$

$$+ \left(\int_{S^{(e)}} [N^{(e)}]^T \left(K_r \frac{\partial \phi}{\partial r} l_r + K_z \frac{\partial \phi}{\partial z} l_z \right) dS \right)$$

$$+ \int_{V^{(e)}} Q^{(e)} [N^{(e)}]^T dV \qquad [6.119]$$

This is the same format as Eqs 6.31 and 6.33 derived for the Cartesian system. With the boundary conditions applied in the same way as before, the weighted residual integral is ultimately found to be

$$- \{G^{(e)}\} = \left(\int_{V^{(e)}} [B^{(e)}]^T [D^{(e)}][B^{(e)}] dV \right.$$

$$+ \int_{S_3^{(e)}} h[N^{(e)}]^T [N^{(e)}] dS \right) \{\Phi^{(e)}\} - \int_{V^{(e)}} Q^{(e)}[N^{(e)}]^T dV$$

$$+ \int_{S_2^{(e)}} q^{(e)}[N^{(e)}]^T dS - \int_{S_3^{(e)}} h\phi_\infty [N^{(e)}]^T dS \qquad [6.120]$$

where

$$\begin{Bmatrix} \dfrac{\partial \phi}{\partial r} \\[2mm] \dfrac{\partial \phi}{\partial z} \end{Bmatrix} = \{g^{(e)}\} = [B^{(e)}]\{\Phi^{(e)}\} \qquad [6.121]$$

$$[D^{(e)}] = \begin{bmatrix} K_r & 0 \\ 0 & K_z \end{bmatrix} \qquad [6.122]$$

with the other terms having their usual meanings.

Thus the governing equations for axisymmetric finite element problems are found to be the same as those for the Cartesian system (Eq. 6.41). However, when the equations are tailored specifically to an axisymmetric simplex element, they prove to be different to the basic two-dimensional simplex element considered previously in this chapter. Since the element assumes a linear variation of temperature in the radial and axial directions,

$$\phi = \alpha_1 + \alpha_2 r + \alpha_3 z \qquad [6.123]$$

This can be expressed in terms of the shape functions in the usual way as

$$\phi = N_i \phi_i + N_j \phi_j + N_k \phi_k = [N]\{\Phi\} \qquad [6.124]$$

where, for example, $N_i = (a_i + b_i r + c_i z)/2A$.

The geometry of the element can be defined in a similar way, in terms of the coordinates of the nodes of the element:

$$
\begin{Bmatrix} r \\ z \end{Bmatrix} = \begin{bmatrix} N_i & 0 & N_j & 0 & N_k & 0 \\ 0 & N_i & 0 & N_j & 0 & N_k \end{bmatrix} \begin{Bmatrix} r_i \\ z_i \\ r_j \\ z_j \\ r_k \\ z_k \end{Bmatrix}
\qquad [6.125]
$$

where (r_i, z_i) are the coordinates of node i, for example.

The $[B]$ matrix is easily found to equal

$$
[B] = \frac{1}{2A} \begin{bmatrix} b_i & b_j & b_k \\ c_i & c_j & c_k \end{bmatrix}
\qquad [6.126]
$$

Therefore the term in the stiffness matrix due to conduction is calculated using Eqs 6.122 and 6.126 (dropping the superscript (e)) as

$$
\int_V [B]^{\mathrm{T}}[D][B]\mathrm{d}V = [B]^{\mathrm{T}} \begin{bmatrix} K_r & 0 \\ 0 & K_z \end{bmatrix} [B]2\pi \int_A r \, \mathrm{d}A
$$

Since $[B]$ and $[D]$ are both constant, they are removed from the integration, and the elemental volume is replaced with $\mathrm{d}V = 2\pi r \, \mathrm{d}A$. The radial position r may be replaced using Eq. 6.125, so that the integral becomes

$$
\dots \int_A r \, \mathrm{d}A = \dots \int_A (N_i r_i + N_j r_j + N_k r_k)\mathrm{d}A
$$

This is easily evaluated using the local integration formula of Eq. 4.26 to give

$$
\frac{A}{3}(r_i + r_j + r_k) = A\underline{r}
$$

where $\underline{r}$ is the radial position of the centroid of the element. Finally, then, the term in the stiffness matrix due to conduction is

$$
2\pi\underline{r}A[B]^{\mathrm{T}} \begin{bmatrix} K_r & 0 \\ 0 & K_z \end{bmatrix} [B]
\qquad [6.127]
$$

When convection occurs from the element, another term must be included in the stiffness matrix:

$$
\int_S h[N]^{\mathrm{T}}[N]\mathrm{d}S
$$

Here S is the surface area over which the convection takes place. For the axisymmetric element in Fig. 6.8, if convection occurs on one side then $\mathrm{d}S$ must be replaced by $\mathrm{d}S = 2\pi r \, \mathrm{d}H$, where H is the distance between nodes. Hence, if convection occurs from side ij, the integral becomes

$$\int_{H_{ij}} h[N]^{\mathrm{T}}[N]2\pi r \, \mathrm{d}H = \int_{H_{ij}} 2\pi h \begin{bmatrix} N_i^2 & N_iN_j & N_iN_k \\ N_jN_i & N_j^2 & N_jN_k \\ N_kN_i & N_kN_j & N_k^2 \end{bmatrix} r \, \mathrm{d}H$$

[6.128]

However, along side ij, $N_k = 0$, so that this equation can be simplified to

$$\int_{H_{ij}} 2\pi h \begin{bmatrix} rN_i^2 & rN_iN_j & 0 \\ rN_jN_i & rN_j^2 & 0 \\ 0 & 0 & 0 \end{bmatrix} \mathrm{d}H$$

If the substitution for r is made again using Eq. 6.125, then the local integration formula of Eq. 4.20 can be used to evaluate the integral directly. The final result is

$$\frac{\pi h H_{ij}}{6} \begin{bmatrix} 3r_i + "r_j & r_i + r_j & 0 \\ r_j + r_i & r_i + 3r_j & 0 \\ 0 & 0 & 0 \end{bmatrix}$$

[6.129]

The convection will also lead to a term in the force vector:

$$\int h\phi_\infty[N]^{\mathrm{T}}\mathrm{d}S = \int_{H_{ij}} h\phi_\infty[N]^{\mathrm{T}}2\pi r \, \mathrm{d}H$$

[6.130]

Replacing r and integrating leads to

$$\frac{\pi h\phi_\infty H_{ij}}{3} \left\{ \begin{array}{c} 2r_i + r_j \\ r_i + 2r_j \\ 0 \end{array} \right\}$$

[6.131]

The contribution to the force vector due to any incident heat flux will result in a similar term:

$$\frac{\pi q H_{ij}}{3} \left\{ \begin{array}{c} 2r_i + r_j \\ r_i + 2r_j \\ 0 \end{array} \right\}$$

[6.132]

Finally, the force vector term due to any internal heat source or sink is calculated from

$$\int_V Q[N]^{\mathrm{T}}\mathrm{d}V = \int_A Q[N]^{\mathrm{T}}2\pi r \, \mathrm{d}A = \int_A 2\pi Q \left\{ \begin{array}{c} rN_i \\ rN_j \\ rN_k \end{array} \right\} \mathrm{d}A$$

Substituting for r and integrating gives

$$\frac{\pi Q A}{6} \left\{ \begin{array}{c} 2r_i + r_j + r_k \\ r_i + 2r_j + r_k \\ r_i + r_j + 2r_k \end{array} \right\}$$

[6.133]

Note that the equivalent nodal forces due to the convection, heat flux and heat source are not distributed equally between the nodes. The node farthest away from the axis of rotation receives the greatest share.

In summary, for an axisymmetric element with convection and heat flux effects on side ij, the following element equations are produced for the stiffness matrix:

$$\left(2\pi \underline{r} A [B]^{\mathrm{T}} \begin{bmatrix} K_r & 0 \\ 0 & K_z \end{bmatrix} [B] + \frac{\pi h H_{ij}}{6} \begin{bmatrix} 3r_i + r_j & r_i + r_j & 0 \\ r_j + r_i & r_i + 3r_j & 0 \\ 0 & 0 & 0 \end{bmatrix} \right)$$

$$\underset{\text{from}}{\text{conduction}} \qquad \underset{\text{from convection}}{\text{on side } ij}$$

$$[6.134]$$

and for the force vector:

$$\left(\frac{\pi h \phi_\infty H_{ij}}{3} \begin{Bmatrix} 2r_i + r_j \\ r_i + 2r_j \\ 0 \end{Bmatrix} + \frac{\pi q H_{ij}}{3} \begin{Bmatrix} 2r_j + r_j \\ r_i + 2r_j \\ 0 \end{Bmatrix} \right.$$

$$\underset{\text{at side } ij}{\text{from convection}} \qquad \underset{\text{on side } ij}{\text{from heat flux}}$$

$$\left. + \frac{\pi Q A}{6} \begin{Bmatrix} 2r_i + r_j + r_k \\ r_i + 2r_j + r_k \\ r_i + r_j + 2r_k \end{Bmatrix} \right) \qquad [6.135]$$

$$\underset{\text{heat source}}{\text{from any internal}}$$

Example 6.6: axisymmetric heat transfer
Calculate the element equations for the axisymmetric element shown in Fig. 6.9, which has an internal heat source and experiences convection on side jk.

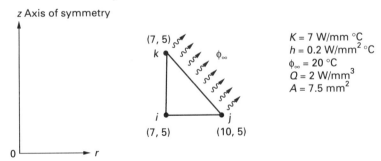

Figure 6.9

The stiffness matrix consists of two terms, one from the conduction, and one from the convection on side jk. Firstly, we need to calculate the $[B]$ matrix, which is given in Eq. 6.126 in terms of the b and c constants (defined in Section 4.6),

$$
\begin{aligned}
b_i &= z_j - z_k = -5 & c_i &= r_k - r_j = -3 \\
b_j &= z_k - z_i = 5 & c_j &= r_i - r_k = 0 \\
b_k &= z_i - z_j = 0 & c_k &= r_j - r_i = 3
\end{aligned}
$$

Since the area of the element is 7.5 mm^2, the $[B]$ matrix is

$$
\frac{1}{15}
\begin{bmatrix}
-5 & 5 & 0 \\
-3 & 0 & 3
\end{bmatrix}
$$

Therefore

$$
2\pi \underline{r} A[B]^{\mathrm{T}} K[B] = \frac{2\pi \underline{r} A}{2A}
\begin{bmatrix}
0.151 & -0.111 & -0.040 \\
-0.111 & 0.111 & 0.0 \\
-0.040 & 0.0 & 0.040
\end{bmatrix}
$$

The radial position of the centroid is 8 mm. Hence the conduction matrix term is

$$
\begin{bmatrix}
398.48 & -292.92 & -105.56 \\
-292.92 & 292.92 & 0.0 \\
-105.56 & 0.0 & 105.56
\end{bmatrix} \tag{6.136}
$$

The convection effect according to Eq. 6.129 (reordered for side jk) produces

$$
\frac{\pi h H_{jk}}{6}
\begin{bmatrix}
0 & 0 & 0 \\
0 & 3r_j + r_k & r_j + r_k \\
0 & r_k + r_j & r_j + 3r_k
\end{bmatrix}
=
\begin{bmatrix}
0 & 0 & 0 \\
0 & 22.59 & 10.38 \\
0 & 10.38 & 18.93
\end{bmatrix} \tag{6.137}
$$

Therefore the total stiffness matrix is the sum of Eqs 6.136 and 6.137, *i.e.*

$$
\begin{bmatrix}
398.48 & -292.92 & -105.56 \\
-292.92 & 315.51 & 10.38 \\
-105.56 & 10.38 & 124.49
\end{bmatrix} \tag{6.138}
$$

The force vector has two terms. From the convection,

$$
\frac{\pi h \phi_\infty H_{jk}}{3}
\begin{Bmatrix}
0 \\
2r_j + r_k \\
r_j + 2r_k
\end{Bmatrix}
=
\begin{Bmatrix}
0 \\
659.47 \\
586.20
\end{Bmatrix} \tag{6.139}
$$

and from the internal heat source,

$$
\frac{\pi Q A}{6}
\begin{Bmatrix}
2r_i + r_j + r_k \\
r_i + 2r_j + r_k \\
r_i + r_j + 2r_k
\end{Bmatrix}
=
\begin{Bmatrix}
73.04 \\
80.11 \\
73.04
\end{Bmatrix} \tag{6.140}
$$

The total force vector is then given by Eqs 6.139 and 6.140:

$$\begin{Bmatrix} 73.04 \\ 739.58 \\ 659.24 \end{Bmatrix} \tag{6.141}$$

So the finite element equations for the specified element must be

$$\begin{bmatrix} 398.48 & -292.92 & -105.56 \\ -292.92 & 315.51 & 10.38 \\ -105.56 & 10.38 & 124.49 \end{bmatrix} \begin{Bmatrix} \phi_i \\ \phi_j \\ \phi_k \end{Bmatrix} = \begin{Bmatrix} 73.04 \\ 739.58 \\ 659.24 \end{Bmatrix}$$

6.4   Torsion problems

Consider a solid prismatic shaft of arbitrary cross-section as shown in Fig. 6.10. All the stresses except τ_{zy} and τ_{zx} are assumed to be zero when the shaft is loaded in torsion. Therefore

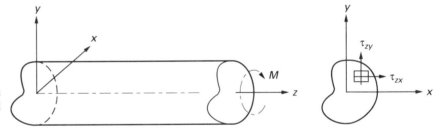

Figure 6.10 Shaft under torsion and the shear stresses produced in the shaft

$$\sigma_{xx} = \sigma_{yy} = \sigma_{zz} = \tau_{xy} = 0 \tag{6.142}$$

The best analytical method of deriving the stress distribution in the shaft is by the use of Prandtl's stress function ϕ, which is defined so that

$$\tau_{zx} = \frac{\partial \phi}{\partial y} \quad \text{and} \quad \tau_{zy} = -\frac{\partial \phi}{\partial x} \tag{6.143}$$

where ϕ can be thought of as a surface covering the cross-section of the shaft, as in Fig. 6.11.

Examination of the shear strains from the displacements leads to the governing differential equation for the shaft of

$$\frac{\partial^2 \phi}{\partial x^2} + \frac{\partial^2 \phi}{\partial y^2} = -2G\theta_1 \tag{6.144}$$

where G is the shear modulus of the material and θ_1 is the twist per unit length of the shaft due to the applied torque.

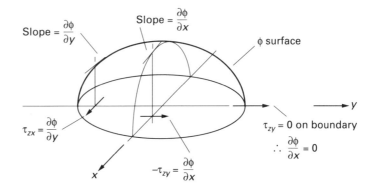

Figure 6.11 The ϕ surface and the shear stresses related to its slope

Since the shear stresses normal to the boundary of the shaft must equal zero, the slope of the surface at that point must equal zero and therefore the value of ϕ must be a constant around the boundary (Fig. 6.11). To simplify the procedures, the following boundary condition is chosen:

$$\phi = 0 \qquad\qquad [6.145]$$

The torque acting on the shaft cross-section may be found by summing the shear stresses at any section, and is easily calculated to be

$$M = 2 \int_A \phi \, \mathrm{d}A \qquad\qquad [6.146]$$

where A is the cross-sectional area of the shaft.

The governing differential equation Eq. 6.144 and the boundary condition Eq. 6.145 are clearly forms of the general field equation Eq. 6.1 and the general boundary conditions Eqs 6.2 to 6.4. The coefficients $K_x = K_y = 1$ and $Q = -2G\theta_1$, while $K_z = q = h = 0$. Therefore the finite element system equations previously derived from the general field equations can be applied directly to the solution of torsion problems, if the coefficients are adjusted accordingly.

Referring to the derivation of the system equations in Section 6.2, note the following for torsion problems. The gradient matrix defined in Eqs 6.9 and 6.15 is

$$\{g\} = \left\{ \begin{array}{c} \dfrac{\partial\phi}{\partial x} \\[2mm] \dfrac{\partial\phi}{\partial y} \end{array} \right\} = \left\{ \begin{array}{c} -\tau_{zy} \\[1mm] \tau_{zx} \end{array} \right\} \qquad\qquad [6.147]$$

and the $[D]$ matrix in Eq. 6.10 is

$$[D] = \begin{bmatrix} 1 & 0 \\ 0 & 1 \end{bmatrix} \qquad\qquad [6.148]$$

The element characteristic equations for torsion problems can then be produced in the same way as Eq. 6.17, and can be shown to be

$$\sum_{e=1}^{E} \left(\int_{V^{(e)}} [B^{(e)}]^{\mathrm{T}}[B^{(e)}]\mathrm{d}V\{\Phi^{(e)}\} - \int_{V^{(e)}} 2G\theta_1[N^{(e)}]^{\mathrm{T}}\mathrm{d}V \right) = 0$$

[6.149]

or

$$\sum_{e=1}^{E} \left([k^{(e)}]\{\Phi^{(e)}\} - \{F^{(e)}\} \right) = 0$$

The torque transmitted is found from Eq. 6.146, which is equivalent to

$$M = \sum_{e=1}^{E} 2 \int_{A^{(e)}} \phi^{(e)} \, \mathrm{d}A$$

[6.150]

if the cross-section of the shaft is modelled with E elements.

Therefore the analysis of torsion problems can be undertaken by the finite element method, solving for the stress function ϕ by the following steps:

(a)   division of the shaft's cross-section into elements
(b)   calculation of each of the element's equations using Eq. 6.149
(c)   assembly of the element equations to produce the system equations
(d)   incorporation of the boundary conditions, and solution of the system equations to find the distribution of the ϕ stress function
(e)   calculation of the shear stresses in each element by Eq. 6.147
(f)   calculation of the torque by Eq. 6.150.

Note that the shear stresses are found from the slope of the ϕ surface, so that if a linear distribution of ϕ is assumed over each element (by the use of simplex elements) the method will predict constant values of stress over each element. Therefore, if a rapid change in stress is expected in the shaft, many elements should be used in that area if the variation is to be correctly modelled.

Example 6.7: a square shaft under torsion

Find the stress distribution in a 20×20 mm^2 square shaft when it is twisted through an angle of $0.5°$ over a length of 1000 mm. Assume the material has a shear modulus of 80×10^3 MPa.

Since a square has four axes of symmetry, only one-eighth of the cross-section needs to be modelled. The meshing chosen is shown in

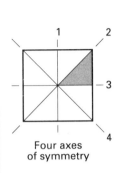
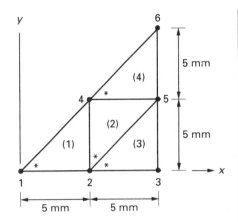

Four axes
of symmetry

Figure 6.12

Fig. 6.12, where the node in an element labelled with a * is the i node, and the other nodes are defined in an anticlockwise manner as usual.

Consider element (1). The distribution of ϕ in the element is given by

$$\phi^{(1)} = N_1\phi_1 + N_2\phi_2 + N_4\phi_4 \qquad [6.151]$$

where

$$\begin{Bmatrix} N_1 \\ N_2 \\ N_4 \end{Bmatrix} = \frac{1}{2A} \begin{bmatrix} a_1 & b_1 & c_1 \\ a_2 & b_2 & c_2 \\ a_4 & b_4 & c_4 \end{bmatrix} \begin{Bmatrix} 1 \\ x \\ y \end{Bmatrix} \qquad [6.152]$$

Therefore from Eq. 6.147 the $[B]$ matrix is calculated to be

$$[B^{(1)}] = \frac{1}{2A} \begin{bmatrix} b_1 & b_2 & b_4 \\ c_1 & c_2 & c_4 \end{bmatrix} \qquad [6.153]$$

The a, b and c constants are defined in Eq. 4.8 and are given by

$$\begin{aligned}
b_1 &= y_2 - y_4 = -5.0 & c_1 &= x_4 - x_2 = 0.0 \\
b_2 &= y_4 - y_1 = 5.0 & c_2 &= x_1 - x_4 = -5.0 \\
b_4 &= y_1 - y_2 = 0.0 & c_4 &= x_2 - x_1 = 5.0
\end{aligned}$$

The area A is 12.5 mm^2, so that

$$[B^{(1)}] = 0.2 \begin{bmatrix} -1 & 1 & 0 \\ 0 & -1 & 1 \end{bmatrix} \qquad [6.154]$$

The stiffness matrix is found from Eq. 6.149 to be

$$[k^{(e)}] = \int_{V^{(e)}} [B^{(e)}]^{\mathrm{T}}[B^{(e)}]\mathrm{d}V = [B^{(e)}]^{\mathrm{T}}[B^{(e)}]A$$

assuming unit thickness. Therefore

$$[k^{(1)}] = \begin{bmatrix} 0.5 & -0.5 & 0 \\ -0.5 & 1.0 & -0.5 \\ 0 & -0.5 & 0.5 \end{bmatrix}$$

The force vector from Eq. 6.149 is

$$\int_{V^{(e)}} 2G\theta_1[N^{(e)}]^T dV = 2G\theta_1 \int_{A^{(e)}} \begin{Bmatrix} N_i \\ N_j \\ N_k \end{Bmatrix} dA$$

which is easily integrated with natural area coordinates to give

$$\{F^{(e)}\} = \frac{2G\theta_1 A^{(e)}}{3} \begin{Bmatrix} 1 \\ 1 \\ 1 \end{Bmatrix} \qquad [6.155]$$

The shaft twists 0.5° over 1000 mm or 8.727×10^{-6} rad/mm and G is given, so that for element (1)

$$\{F^{(e)}\} = 5.818 \begin{Bmatrix} 1 \\ 1 \\ 1 \end{Bmatrix} \qquad [6.156]$$

The other element equations can be calculated in the same way and are found to be as follows

$$[k^{(2)}] = \begin{bmatrix} 0.5 & 0.0 & -0.5 \\ 0.0 & 0.5 & -0.5 \\ -0.5 & -0.5 & 1.0 \end{bmatrix} \qquad \{F^{(2)}\} = 5.818 \begin{Bmatrix} 1 \\ 1 \\ 1 \end{Bmatrix}$$

$$[k^{(3)}] = \begin{bmatrix} 0.5 & -0.5 & 0.0 \\ -0.5 & 1.0 & -0.5 \\ 0.0 & -0.5 & 0.5 \end{bmatrix} \qquad \{F^{(3)}\} = 5.818 \begin{Bmatrix} 1 \\ 1 \\ 1 \end{Bmatrix}$$

$$[k^{(4)}] = \begin{bmatrix} 0.5 & -0.5 & 0.0 \\ -0.5 & 1.0 & -0.5 \\ 0.0 & -0.5 & 0.5 \end{bmatrix} \qquad \{F^{(4)}\} = 5.818 \begin{Bmatrix} 1 \\ 1 \\ 1 \end{Bmatrix}$$

Assembly of these equations then gives the system equations as follows:

$$\begin{bmatrix} 0.5 & -0.5 & 0.0 & 0.0 & 0.0 & 0.0 \\ -0.5 & 2.0 & -0.5 & -1.0 & 0.0 & 0.0 \\ 0.0 & -0.5 & 1.0 & 0.0 & -0.5 & 0.0 \\ 0.0 & -1.0 & 0.0 & 2.0 & -1.0 & 0.0 \\ 0.0 & 0.0 & -0.5 & -1.0 & 2.0 & -0.5 \\ 0.0 & 0.0 & 0.0 & 0.0 & -0.5 & 0.5 \end{bmatrix} \begin{Bmatrix} \phi_1 \\ \phi_2 \\ \phi_3 \\ \phi_4 \\ \phi_5 \\ \phi_6 \end{Bmatrix} = \begin{Bmatrix} 5.818 \\ 17.454 \\ 5.818 \\ 17.454 \\ 17.454 \\ 5.818 \end{Bmatrix}$$

$$[6.157]$$

The boundary conditions for the problem are now substituted into these equations, namely that the values of ϕ_3, ϕ_5 and ϕ_6 must be zero, since they are on the surface of the shaft. Solution of the equations then gives

$$\{\Phi_S\}^T = [43.635\ 31.999\ 0\ 24.726\ 0\ 0] \qquad [6.158]$$

To calculate the shear stresses, use Eq. 6.147 for each element with the value of the $[B]$ matrix previously calculated in the derivation of the element's stiffness matrix. For example, for element (1)

$$\{g^{(1)}\} = [B^{(1)}]\{\Phi^{(1)}\} = \begin{bmatrix} -0.2 & 0.2 & 0.0 \\ 0.0 & -0.2 & 0.2 \end{bmatrix} \begin{Bmatrix} \phi_1 \\ \phi_2 \\ \phi_4 \end{Bmatrix}$$

or

$$\begin{Bmatrix} \tau_{zy} \\ \tau_{zx} \end{Bmatrix} = \begin{Bmatrix} 2.33 \\ -1.45 \end{Bmatrix} \text{N/mm}^2$$

The other elements yield the following:

Element	Shear stresses (N/mm^2)	
	τ_{zy}	τ_{zx}
2	4.95	−1.45
3	6.40	0.0
4	4.95	0.0

Thus the finite element method predicts the stress pattern shown in Fig. 6.13 for the square shaft under torsion. The stresses are constant over each element, and so, with just these four elements, little detail can be obtained from the results, as plotted on Fig. 6.14. Extra processing and interpretation of results of finite element models can be performed with care, and is discussed in detail in Chapter 9. In fact to obtain a little more information from the

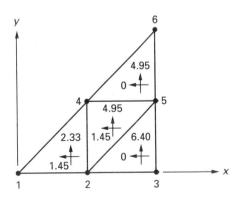

Figure 6.13 Shear stresses in the model (N/mm^2)

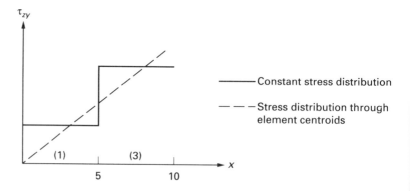

Figure 6.14 Shear stress distribution along axis x

results of τ_{zy} it is not unreasonable to take the calculated values as those at the centroids of the elements (since they are constant stress elements). A line drawn through these points then produces a stress pattern more in keeping with the distribution than would be expected (*i.e.* $\tau_{zy} = 0$ at the centre, and a maximum on the surface of the shaft).

The theoretical solution for this problem predicts a maximum shear stress of 9.43 N/mm². This compares with 6.40 N/mm² for the constant stress distribution (32.1 per cent error), and 7.76 N/mm² from the centroidal values (17.6 per cent error).

The torque required to produce the twist of 0.5° per 1000 mm length is found from Eq. 6.150. For any element, since the nodal values of the stress function are known,

$$M = 2 \int_{A^{(e)}} \phi^{(e)} dA = 2 \int_{A^{(e)}} \{\Phi^{(e)}\}^T [N^{(e)}]^T dA$$

$$= 2\{\Phi^{(e)}\}^T \int_{A^{(e)}} [N^{(e)}]^T dA$$

This is evaluated using natural coordinate integration formulae, so that

$$M^{(e)} = \frac{2A\{\Phi^{(e)}\}^T}{3} \begin{Bmatrix} 1 \\ 1 \\ 1 \end{Bmatrix} \qquad [6.159]$$

For element (1),

$$M^{(1)} = \frac{2A}{3} (43.635 \quad 31.999 \quad 24.726) \begin{Bmatrix} 1 \\ 1 \\ 1 \end{Bmatrix} = 836.3 \text{ N mm}$$

The process is repeated for the other elements, to show

$$M^{(2)} = 472.7 \qquad M^{(3)} = 266.6 \qquad M^{(4)} = 206.1 \text{ N mm}$$

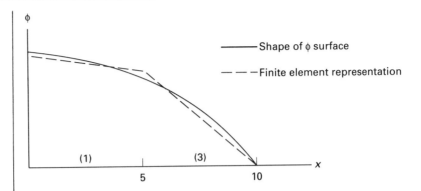

Figure 6.15

Therefore the torque on the part of the shaft that is modelled is 1781.7 N mm, and so the total torque that must act on the complete shaft to produce the given twist is

$$8 \times 1781.7 = 14\ 258.6\ \text{N mm}$$

Theory predicts the required torque will be 15 708 N mm. The finite element method underestimates the torque by 9.3 per cent.

The meshing of the shaft cross-section is very coarse, and uses very few elements. It is surprising therefore that the results are not more inaccurate. In such a case, it is fortunate that the linear interpolation function used models the shape of the ϕ surface so closely.

The variation of ϕ across the shaft is approximately quadratic, and the elements used here model the distribution as illustrated in Fig. 6.15. The finite element representation of the surface has straight sides, and therefore predicts a constant value of shear stress in the two regions shown, as previously discovered. The torque carried by the shaft is represented by the volume under the ϕ surface, so it can be seen that the area, and hence the volume, under the modelled ϕ surface is less than the actual, and hence the torque is less.

In practice, to evaluate the stresses sensibly, the shape of the ϕ surface would be modelled more accurately, either by using more elements and replacing the curved shape by more straight sections, or by using higher-order elements which can take up quadratic, cubic and higher-order shapes.

Examples of the use of the finite element method to predict the shape of various ϕ surfaces of several cross-sections are included in Plate 1. Remember that the shear stresses are obtained from the slope of the surface at any point, and the torque carried is proportional to the volume under the surface.

6.5   Fluid flow problems

The application of the finite element method to fluid flow problems is not as advanced as it is in structural and solid mechanics or thermal problems. The reason is not that the method is unsuitable, but rather that the finite difference method usually used for fluid flow problems is so successful. Thus there has been no real motivation to extend the finite element method to fluid problems, and the large investments of time and money in the development of sophisticated finite difference software have naturally led to a further reluctance to consider other methods. This state of affairs is however changing as more sophisticated modelling situations are encountered, and the finite element method is now contributing to the solution of more fluid flow problems.

The range of problems that are being investigated is large, for example:

(a)   inviscid incompressible flow: flow around corners, over cylinders and aerofoils, and through nozzles etc.
(b)   flow in porous media: flow through and under dams, retaining walls and foundations; aquifer analysis; flow towards ditches, wells and canals
(c)   wave motion in shallow basins or lakes: for the design of harbours etc.
(d)   incompressible viscous flow
(e)   flow of non-Newtonian fluids: for example, crude oil, slurries and suspensions.

As an introductory text on the finite element method, this book does not cover these areas, and any potential user is referred to other more advanced and specialized texts for further information. The topics are listed here, since it is important that the reader is at least aware of what the finite element method can do.

6.6   Conclusions

The number of engineering problems governed by the field equation (or a modified form of it) is extensive, and these problems cannot all be covered in an introductory text. The fundamental finite element equations are derived in this chapter for the field equation, and are then applied to thermal problems and the torsion of shafts by adjusting the coefficients as necessary. The same approach is used for the other situations that are not discussed here. The procedure is straightforward, and should not prove difficult if the principles detailed in this chapter are followed.

Throughout the chapter, the sample problems have been analysed

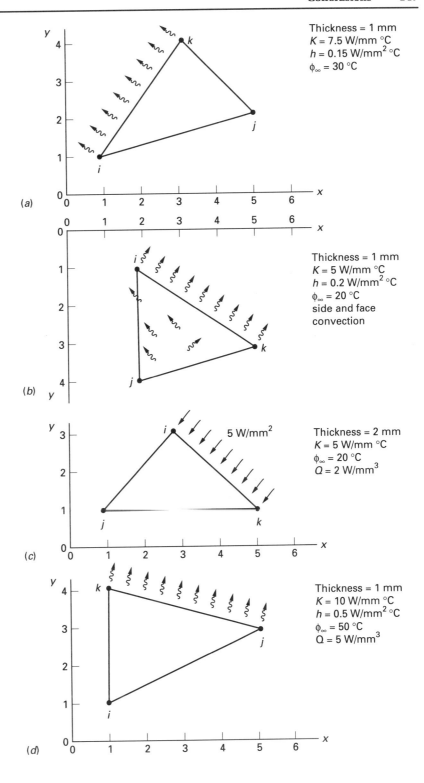

Figure 6.16

using simplex elements. The higher-order elements which are introduced in Chapter 8 can be applied in the same way to the governing equations. The resulting finite element equations have a similar format, but of course are larger in number. For example, a triangular element using a quadratic interpolation function will have three corner and three mid-side nodes, and therefore needs an element conductance matrix of size $[6 \times 6]$ and a force vector of $\{6 \times 1\}$ to describe it. The number of calculations in an analysis using higher-order elements clearly increases rapidly and becomes awkward to perform by hand. However, some of the problems presented in this chapter are re-evaluated in Chapter 8 using higher-order elements. The improvements achieved by the more complex elements are dramatic and important.

Problems

6.1   Reanalyse the fin in Example 6.1 assuming a square cross-section of size 5×5 mm.

6.2   Reanalyse the fin in Example 6.1 assuming a circular cross-section of diameter 10 mm.

6.3   Examine the convergence of the finite element analysis in Example 6.1 by comparing the results predicted by models using one, two, three and four elements.

6.4   Vary the length of the fin in Example 6.1 and compare the efficiencies of the different designs.

6.5   Vary the cross-sectional area of the fin in Example 6.1 and compare the efficiencies of the different designs.

6.6   Model the fin in Example 6.1 with various cross-sectional shapes (*e.g.* circular, square or triangular) while maintaining the same volume of material, and compare the efficiencies of the different designs.

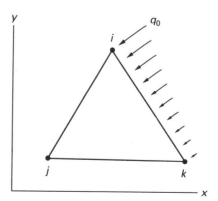

Figure 6.17

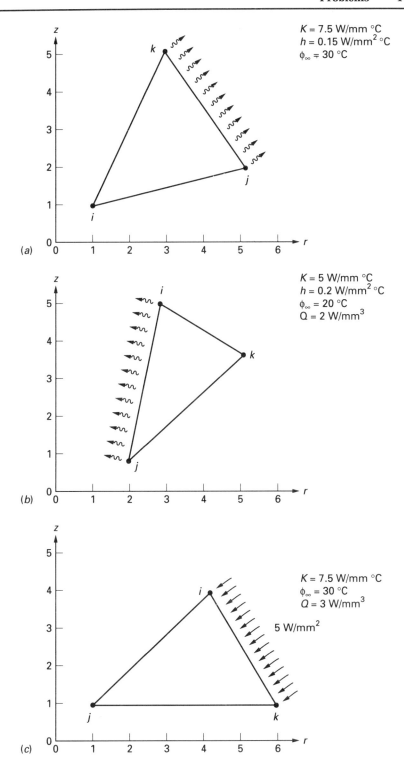

Figure 6.18

6.7   Remodel the fin analysed in Example 6.2 but with the taper in the opposite direction, and compare the performances of the two designs.

6.8   Calculate the finite element equations for the simplex elements shown in Fig. 6.16(a)–(d).

6.9   The calculated temperatures (°C) of the elements in Problem 6.8 are shown in the following table. What are the heat flows in each element?

Element	T_i	T_j	T_k
(a)	16	20	35
(b)	25	15	10
(c)	0	20	32
(d)	5	18	24

6.10   Derive the force vector term required to take account of the linearly varying heat flux shown in Fig. 6.17. [Hint: express the flux distribution as a function of the nodal values using the shape functions.]

6.11   Prove that the terms required in the force vector and stiffness matrix due to convection over side jk of an axisymmetric simplex element are respectively

$$\frac{\pi h \phi_\infty H_{jk}}{3} \begin{Bmatrix} 0 \\ 2r_j + r_k \\ r_j + 2r_k \end{Bmatrix} \quad \text{and} \quad \frac{\pi h H_{jk}}{6} \begin{bmatrix} 0 & 0 & 0 \\ 0 & 3r_j + r_k & r_j + r_k \\ 0 & r_j + r_k & r_j + 3r_k \end{bmatrix}$$

6.12   Calculate the finite element equations for the axisymmetric simplex elements shown in Fig. 6.18(a)–(c).

6.13   Calculate the finite element equations for the simplex torsion elements shown in Fig. 6.19(a) and (b).

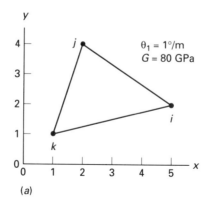

(a)

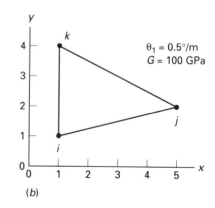

(b)

Figure 6.19

6.14 The calculated values of Prandtl's stress function for the elements in Problem 6.13 are shown in the following table. What are the resultant element stresses and torques carried by each element?

Element	ϕ_i	ϕ_j	ϕ_k
(a)	20	35	15
(b)	10	25	40

7 Assembly and solution of the finite element equations

7.1 Introduction

Earlier chapters of this book have discussed development of the finite element model, the concept of interpolation functions and in particular those of simplex elements, and the derivation of the element equations for elasticity and field problems. The next stages in the finite element procedure are the assembly and solution of the system equations. In fact these ideas were introduced in the two simple examples of Sections 2.2.1 and 2.3.1, which indicates that they are not difficult concepts to understand. From a computer processing point of view, however, they are the most time consuming and demanding part of the analysis. Specifically, the solution of the equations is the most computer intensive phase, and the maximum size of the model that a hardware configuration can analyse is usually determined by this step. For this reason, considerable time and effort have been spent on the development of efficient solution techniques over the years. A detailed discussion of these techniques is not included here because, although they are important, they are not directly relevant to the modelling processes and are generally outside the direct control of the user of commercial finite element programs. It is important, however, that the user should at least understand the basic principles, and know the type of solution method employed by a package, so that the model can be constructed in the most efficient and economic way. Such careful modelling will result in a faster (and cheaper) solution, or the possibility of analysing a larger or more detailed model.

7.2 Coordinate transformations

Before some element stiffness matrices can be assembled into the global stiffness matrix, it is necessary to perform a coordinate transformation on the matrix. This occurs when it is easier to generate the element matrix in a local coordinate system rather than the global system. An example of this type of transformation is used with the pin-jointed bar, introduced in Chapter 5.

The conversion from one coordinate system to another involves a transformation matrix $[\lambda]$, which is used to pre- and post-multiply the stiffness matrix derived in the local coordinate system as follows:

$$[k^\circ] = [\lambda]^\mathrm{T}[k][\lambda] \qquad [7.1]$$

Here $[k^\circ]$ is the element matrix in the global coordinate system, and $[\lambda]$ is defined so that it relates the local displacement $\{U\}$ to the global values $\{U^\circ\}$ by

$$\{U\} = [\lambda]\{U^\circ\} \qquad [7.2]$$

For example, the local stiffness matrix of the pin-jointed bar in Chapter 5 is found to be (Eq. 5.27)

$$[k] = \frac{AE}{L}\begin{bmatrix} 1 & -1 \\ -1 & 1 \end{bmatrix}$$

and the transformation matrix in two-dimensional space is (Eq. 5.47)

$$[\lambda] = \begin{bmatrix} l & m & 0 & 0 \\ 0 & 0 & l & m \end{bmatrix}$$

where l and m are the direction cosines as defined in Fig. 7.1.

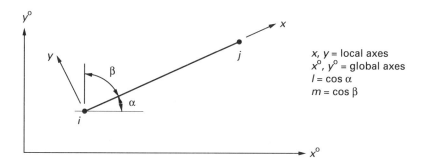

Figure 7.1 Position of the one-dimensional pin-jointed bar in two-dimensional space

When the local stiffness matrix has been transformed using Eq. 7.1, the global matrix is found to be

$$[k^\circ] = \frac{AE}{L}\begin{bmatrix} l^2 & lm & -l^2 & -lm \\ lm & m^2 & -lm & -m^2 \\ -l^2 & -lm & l^2 & lm \\ -lm & -m^2 & lm & m^2 \end{bmatrix}$$

Note that since each of the two nodes has one degree of freedom in one-dimensional space, but two degrees of freedom in two-dimensional space, the size of the matrix increases from a $[2 \times 2]$ to a $[4 \times 4]$ square matrix.

Example 7.1: calculation of a coordinate transformation matrix

A beam element is a one-dimensional element that can resist bending moments applied in its plane (unlike a pin-jointed bar, which can only support axial loads), and it is discussed in detail in Section 8.5. If a cubic interpolation function is assumed for the vertical displacement of the beam, it will have the form

$$v = a_1 + a_2 x + a_3 x^2 + a_4 x^3$$

where x is the local coordinate of the beam shown in Fig. 7.2. The beam has a total of four degrees of freedom; v_i and v_j are vertical translations, while θ_i and θ_j are rotations.

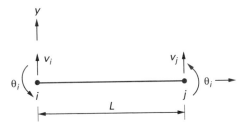

Figure 7.2

The stiffness matrix for the beam in local coordinates is found to be

$$[k] = \frac{EI}{L^3} \begin{bmatrix} 12 & 6L & -12 & 6L \\ 6L & 4L^2 & -6L & 2L^2 \\ -12 & -6L & 12 & -6L \\ 6L & 2L^2 & -6L & 4L^2 \end{bmatrix} \qquad [7.3]$$

where the displacement vector is

$$\{U\}^{\mathrm{T}} = [v_i \ \theta_i \ v_j \ \theta_j] \qquad [7.4]$$

What is the transformation matrix required to use the element in two-dimensional space?

In two dimensions, each node will have three degrees of freedom, *i.e.* two displacements and one rotation. The rotation will be the same whichever coordinate system is being used, but the one-dimensional translation v will need to be transformed into global directions to give u^o and v^o. The transformation matrix will therefore be of the form

$$\begin{Bmatrix} v_i \\ \theta_i \\ v_j \\ \theta_j \end{Bmatrix} = \begin{bmatrix} 6 \times 4 \end{bmatrix} \begin{Bmatrix} u_i^o \\ v_i^o \\ \theta_i \\ u_j^o \\ v_j^o \\ \theta_j^o \end{Bmatrix} = [\lambda]\{U^o\} \qquad [7.5]$$

If the beam element is oriented in space in the same way as the element in Fig. 7.1, then the vertical displacement at node i, for example, is found to be

$$v_i = (-\cos \beta)u_i^o + (\cos \alpha)v_i^o = -mu_i^o + lv_i^o \qquad [7.6]$$

(as shown in Fig. 5.5).

As stated before, the rotations at each node are the same for both systems; hence

$$\theta_i = \theta_i^o \qquad [7.7]$$

Therefore the transformation matrix must be

$$[\lambda] = \begin{bmatrix} -m & l & 0 & 0 & 0 & 0 \\ 0 & 0 & 1 & 0 & 0 & 0 \\ 0 & 0 & 0 & -m & l & 0 \\ 0 & 0 & 0 & 0 & 0 & 1 \end{bmatrix} \qquad [7.8]$$

7.3   Assembly of the element equations

Assembly of the element equations into the system equations is simply a question of adding the coefficients of each element stiffness matrix into the corresponding places of the global stiffness matrix, and summing the force vector coefficients into the global force vector. The procedures are the same regardless of the type of problem and the number and type of elements used. Again this has already been performed in the sample problems of earlier chapters.

The easiest way to assemble the elements is to label each row and column of the element matrix with its corresponding degree of freedom, and then to work through the coefficients of the matrix, adding each into the global matrix which has been similarly labelled.

The following simple example demonstrates the principles. A sample mesh for the analysis of a thermal problem (one degree of freedom per node) is shown in Fig. 7.3. Since the problem has cyclic symmetry, only

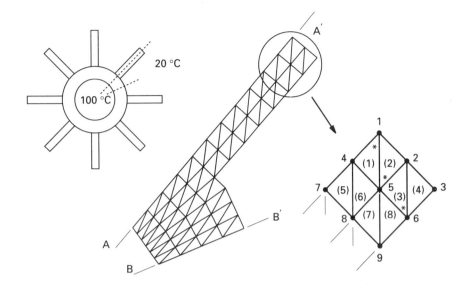

Figure 7.3 Finite element model of a pipe with cooling fins, showing the detail of the element meshing

one-sixteenth of the pipe and fins needs to be modelled as shown, provided the correct boundary conditions are applied. (The heat flow in the circumferential direction is zero along the boundaries AA' and BB'.)

Assume for the sake of this example that the stiffness matrices of the first three elements are

$$[k^{(1)}] = \begin{bmatrix} 7 & 3 & 1 \\ 3 & 6 & 2 \\ 1 & 2 & 5 \end{bmatrix} \qquad [k^{(2)}] = \begin{bmatrix} 8 & 1 & 2 \\ 1 & 7 & 3 \\ 2 & 3 & 4 \end{bmatrix} \qquad [k^{(3)}] = \begin{bmatrix} 9 & 4 & 1 \\ 4 & 6 & 0 \\ 1 & 0 & 5 \end{bmatrix}$$

where the i nodes are identified by an asterisk in Fig. 7.3, and the elements are labelled in an anticlockwise manner.

The total model has 56 nodes and, with one degree of freedom per node, the global stiffness matrix will be [56 × 56]. However, for this example only the first six degrees of freedom will be considered.

Consider element (1). The rows and columns are identified as follows:

$$[k^{(1)}] = \begin{matrix} & 1 \ 4 \ 5 \\ \begin{bmatrix} 7 & 3 & 1 \\ 3 & 6 & 2 \\ 1 & 2 & 5 \end{bmatrix} & \begin{matrix} 1 \\ 4 \\ 5 \end{matrix} \end{matrix}$$

They are in the order 1–4–5 because 1 is the i node, and 4 and 5 follow in an anticlockwise direction. The matrix is added into the (empty) global matrix to give

1 2 3 4 5 6

$$\begin{bmatrix} 7 & 0 & 0 & 3 & 1 & 0 & \dots \\ 0 & 0 & 0 & 0 & 0 & 0 \\ 0 & 0 & 0 & 0 & 0 & 0 \\ 3 & 0 & 0 & 6 & 2 & 0 \\ 1 & 0 & 0 & 2 & 5 & 0 \\ 0 & 0 & 0 & 0 & 0 & 0 \\ & & \cdot & & & \\ & & \cdot & & & \\ & & \cdot & & & \end{bmatrix} \begin{matrix} 1 \\ 2 \\ 3 \\ 4 \\ 5 \\ 6 \end{matrix}$$

The matrix of the second element is labelled as follows:

5 2 1

$$[k^{(2)}] = \begin{bmatrix} 8 & 1 & 2 \\ 1 & 7 & 3 \\ 2 & 3 & 4 \end{bmatrix} \begin{matrix} 5 \\ 2 \\ 1 \end{matrix}$$

This is added into the global matrix as follows to give

1   2   3   4   5   6

$$\begin{bmatrix} (7+4) & 3 & 0 & 3 & (1+2) & 0 & \dots \\ 3 & 7 & 0 & 0 & 1 & 0 \\ 0 & 0 & 0 & 0 & 0 & 0 \\ 3 & 0 & 0 & 6 & 2 & 0 \\ (1+2) & 1 & 0 & 2 & (5+8) & 0 \\ 0 & 0 & 0 & 0 & 0 & 0 \\ & & \cdot & & & \\ & & \cdot & & & \\ & & \cdot & & & \end{bmatrix} \begin{matrix} 1 \\ 2 \\ 3 \\ 4 \\ 5 \\ 6 \end{matrix}$$

For element (3), the stiffness matrix is

6 2 5

$$[k^{(3)}] = \begin{bmatrix} 9 & 4 & 1 \\ 4 & 6 & 0 \\ 1 & 0 & 5 \end{bmatrix} \begin{matrix} 6 \\ 2 \\ 5 \end{matrix}$$

This is added into the global matrix to give

$$
\begin{array}{cccccc}
1 & 2 & 3 & 4 & 5 & 6
\end{array}
$$

$$
\begin{bmatrix}
11 & 3 & 0 & 3 & 3 & 0 & \ldots \\
3 & (7+6) & 0 & 0 & (1+0) & 4 & \\
0 & 0 & 0 & 0 & 0 & 0 & \\
3 & 0 & 0 & 6 & 2 & 0 & \\
3 & (1+0) & 0 & 2 & (13+5) & 1 & \\
0 & 4 & 0 & 0 & 1 & 9 & \\
\vdots & & & & & & \\
\end{bmatrix}
\begin{matrix}
1 \\ 2 \\ 3 \\ 4 \\ 5 \\ 6 \\
\end{matrix}
$$

Note that after adding in details of the first three elements, some of the slots in the global matrix are still equal to zero. The reason for this is that none of the three elements links the associated degrees of freedom. For example, the coefficient in positions (4,2) and (2,4) is zero. Reference to Fig. 7.3 shows that node 4 is common to elements (1), (5) and (6), whereas node 2 only occurs in elements (2), (3) and (4); hence the temperatures of nodes 2 and 4 are not linked.

The addition of the element force vectors into the global force vector proceeds in exactly the same way, except of course that there is only one column to consider, making the task that much easier.

When the finite element method is used to investigate the distribution of a vector quantity, for example in stress analysis problems, the use of the simplified notation first introduced in Section 4.5 makes the development of the system equations as equally straightforward.

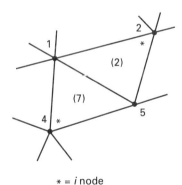

* = *i* node

Elements defined in
anticlockwise direction

Figure 7.4

Example 7.2: assembly of element equations

Assemble the given element matrices and vectors for elements (2) and (7) of the two-dimensional (stress analysis) mesh shown in Fig. 7.4, where

$$
[k^{(2)}] =
\begin{bmatrix}
22 & -3 & -7 & -4 & -6 & -2 \\
-3 & 29 & -9 & -9 & -1 & -7 \\
-7 & -9 & 30 & -6 & -3 & -5 \\
-4 & -9 & -6 & 31 & -4 & -8 \\
-6 & -1 & -3 & -4 & 16 & -2 \\
-2 & -7 & -5 & -8 & -2 & 24
\end{bmatrix}
\qquad
\{F^{(2)}\} =
\begin{Bmatrix}
3 \\ 6 \\ 4 \\ 1 \\ 7 \\ 5
\end{Bmatrix}
$$

$$[k^{(7)}] = \begin{bmatrix} 23 & -1 & -6 & -8 & -3 & -5 \\ -1 & 19 & -2 & -4 & -7 & -5 \\ -6 & -2 & 30 & -7 & -8 & -7 \\ -8 & -4 & -7 & 25 & -2 & -4 \\ -3 & -7 & -8 & -2 & 27 & -7 \\ -5 & -5 & -7 & -4 & -7 & 28 \end{bmatrix} \qquad \{F^{(7)}\} = \begin{Bmatrix} 9 \\ 7 \\ 6 \\ 2 \\ 4 \\ 5 \end{Bmatrix}$$

The nodes of element (2) are 2–1–5. Therefore the corresponding degrees of freedom (as defined in Fig. 4.16) are

$$\begin{Bmatrix} u_{2i-1} \\ u_{2i} \\ u_{2j-1} \\ u_{2j} \\ u_{2k-1} \\ u_{2k} \end{Bmatrix} = \begin{Bmatrix} u_3 \\ u_4 \\ u_1 \\ u_2 \\ u_9 \\ u_{10} \end{Bmatrix}$$

The rows and columns of the stiffness matrix of element (2) must be labelled accordingly, and the terms then placed into the global stiffness matrix:

$$\begin{array}{cccccc} 3 & 4 & 1 & 2 & 9 & 10 \end{array}$$

$$\begin{bmatrix} 22 & -3 & -7 & -4 & -6 & -2 \\ -3 & 29 & -9 & -9 & -1 & -7 \\ -7 & -9 & 30 & -6 & -3 & -5 \\ -4 & -9 & -6 & 31 & -4 & -8 \\ -6 & -1 & -3 & -4 & 16 & -2 \\ -2 & -7 & -5 & -8 & -2 & 24 \end{bmatrix} \begin{array}{c} 3 \\ 4 \\ 1 \\ 2 \\ 9 \\ 10 \end{array}$$

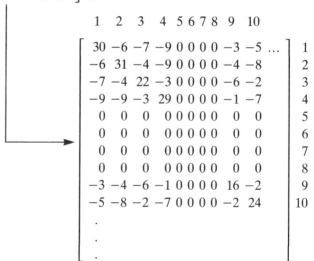

The nodes of element (7) are 4–5–1, and hence the degrees of freedom are (7, 8, 9, 10, 1, 2). Addition into the global stiffness matrix gives

$$
\begin{array}{cccccc}
7 & 8 & 9 & 10 & 1 & 2
\end{array}
$$

$$
\begin{bmatrix}
23 & -1 & -6 & -8 & -3 & -5 \\
-1 & 19 & -2 & -4 & -7 & -5 \\
-6 & -2 & 30 & -7 & -8 & -7 \\
-8 & -4 & -7 & 25 & -2 & -4 \\
-3 & -7 & -8 & -2 & 27 & -7 \\
-5 & -5 & -7 & -4 & -7 & 28
\end{bmatrix}
\begin{array}{c}
7 \\ 8 \\ 9 \\ 10 \\ 1 \\ 2
\end{array}
$$

$$
\begin{array}{cccccccccc}
1 & 2 & 3 & 4 & 5 & 6 & 7 & 8 & 9 & 10
\end{array}
$$

$$
\begin{bmatrix}
57 & -13 & -7 & -9 & 0 & 0 & -3 & -7 & -11 & -7 \ldots \\
-13 & 59 & -4 & -9 & 0 & 0 & -5 & -5 & -11 & -12 \\
-7 & -4 & 22 & -3 & 0 & 0 & 0 & 0 & -6 & -2 \\
-9 & -9 & -3 & 29 & 0 & 0 & 0 & 0 & -1 & -7 \\
0 & 0 & 0 & 0 & 0 & 0 & 0 & 0 & 0 & 0 \\
0 & 0 & 0 & 0 & 0 & 0 & 0 & 0 & 0 & 0 \\
-3 & -5 & 0 & 0 & 0 & 0 & 23 & -1 & -6 & -8 \\
-7 & -5 & 0 & 0 & 0 & 0 & -1 & 19 & -2 & -4 \\
-11 & -11 & -6 & -1 & 0 & 0 & -6 & -2 & 46 & -9 \\
-7 & -12 & -2 & -7 & 0 & 0 & -8 & -4 & -9 & 49
\end{bmatrix}
\begin{array}{c}
1 \\ 2 \\ 3 \\ 4 \\ 5 \\ 6 \\ 7 \\ 8 \\ 9 \\ 10
\end{array}
$$

The force vectors are summed in to the global force vector in a similar way:

$$
\{F^{(2)}\} = \begin{Bmatrix} 3 \\ 6 \\ 4 \\ 1 \\ 7 \\ 5 \end{Bmatrix}
\begin{array}{c} 3 \\ 4 \\ 1 \\ 2 \\ 9 \\ 10 \end{array}
\qquad
\begin{Bmatrix} 4 \\ 1 \\ 3 \\ 6 \\ 0 \\ 0 \\ 0 \\ 0 \\ 7 \\ 5 \end{Bmatrix}
\qquad
\{F^{(7)}\} = \begin{Bmatrix} 9 \\ 7 \\ 6 \\ 2 \\ 4 \\ 5 \end{Bmatrix}
\begin{array}{c} 7 \\ 8 \\ 9 \\ 10 \\ 1 \\ 2 \end{array}
\qquad
\begin{Bmatrix} 8 \\ 6 \\ 3 \\ 6 \\ 0 \\ 0 \\ 9 \\ 7 \\ 13 \\ 7 \end{Bmatrix}
$$

This completes the assembly of the two element equations.

7.4   Incorporation of the boundary conditions

Once the element equations have been assembled to give the system equations, the boundary conditions of the problem must be incorporated. The equations cannot be solved without applying any boundary conditions because the stiffness matrix will be singular, and hence its inverse will not exist. For stress analysis problems, this physically means that the structure will undergo unlimited rigid body motion, unless restraining forces are applied to keep the model in equilibrium. Note that sufficient constraints must be applied to completely restrain the model. Constraining the displacements of one node in three directions is not enough for a full three-dimensional analysis, because the body can still rotate about that point. Either the node's rotations must be suppressed, or other nodes must be constrained as well. Failing to adequately restrain a model is a common mistake.

As with many aspects of the finite element method, there are several ways in which the boundary conditions can be incorporated into the system equations. One method is to rearrange the equations and to partition the matrix so that all the specified degrees of freedom are together, *i.e.*

$$\begin{bmatrix} [k_{11}] & [k_{12}] \\ [k_{21}] & [k_{22}] \end{bmatrix} \begin{Bmatrix} \{U_1\} \\ \{U_2\} \end{Bmatrix} = \begin{Bmatrix} \{F_1\} \\ \{F_2\} \end{Bmatrix} \tag{7.9}$$

where $\{U_1\}$ is the vector of unknown degrees of freedom, while those in $\{U_2\}$ are all specified. Consequently, $\{F_1\}$ will contain only the known nodal forces, and $\{F_2\}$ will contain the unknown reactions.

Equation 7.9 can be written as two equations:

$$[k_{11}]\{U_1\} + [k_{12}]\{U_2\} = \{F_1\} \tag{7.10}$$
$$[k_{21}]\{U_1\} + [k_{22}]\{U_2\} = \{F_2\} \tag{7.11}$$

Then the first of these can be rearranged in the form

$$[k_{11}]\{U_1\} = \{F_1\} - [k_{12}]\{U_2\} = \{F_1^*\} \tag{7.12}$$

Since $\{U_2\}$ is known, the terms on the right-hand side can be reduced to a simple vector $\{F_1^*\}$, and the resulting equation can then be solved in a standard way for the unknown variables $\{U_1\}$.

Now that the unknown degrees of freedom have been calculated, Eq. 7.11 is used to calculate the reactions $\{F_2\}$.

This method of dealing with the defined boundary conditions is straightforward, but it does require the equations to be renumbered, since it is most unlikely that the specified degrees of freedom will occur at the end of the vector $\{U\}$. The following method is similar, but it does not require the equations to be reordered. To understand the method, firstly consider Eq. 7.9. This can be rewritten using Eq. 7.12 as

$$\begin{bmatrix} [k_{11}] & [0] \\ [k_{21}] & [k_{22}] \end{bmatrix} \begin{Bmatrix} \{U_1\} \\ \{U_2\} \end{Bmatrix} = \begin{Bmatrix} \{F_1\} - [k_{12}]\{U_2\} \\ \{F_2\} \end{Bmatrix} = \begin{Bmatrix} \{F_1^*\} \\ \{F_2\} \end{Bmatrix}$$
[7.13]

If the equations in the second line of the above matrix are stored in a temporary matrix for later use, then Eq. 7.13 can be written as

$$\begin{bmatrix} [k_{11}] & [0] \\ [0] & [k_{22}^*] \end{bmatrix} \begin{Bmatrix} \{U_1\} \\ \{U_2\} \end{Bmatrix} = \begin{Bmatrix} \{F_1^*\} \\ \{F_2^*\} \end{Bmatrix}$$
[7.14]

where $[k_{22}^*]$ is the matrix $[k_{22}]$ with all the off-diagonal terms set equal to zero, and $\{F_2^*\}$ is the correct product of $[k_{22}^*]$ and $\{U_2\}$. The second line is of no use, but it does allow the matrix and vectors to remain the same size, and in fact the above process can be performed without reordering the equations. Once the degrees of freedom are evaluated by solving Eq. 7.14, the reactions can be calculated from the temporary matrix, and the solution is complete. The following example demonstrates the method:

Example 7.3: incorporation of boundary conditions

Incorporate the specified boundary conditions into the following system of equations:

$$\begin{bmatrix} 44 & -12 & 3 & -5 & 0 & 0 \\ -12 & 58 & -4 & 10 & 6 & 0 \\ 3 & -4 & 32 & 7 & 3 & -1 \\ -5 & 10 & 7 & 48 & 2 & 11 \\ 0 & 6 & 3 & 2 & 60 & 20 \\ 0 & 0 & -1 & 11 & 20 & 62 \end{bmatrix} \begin{Bmatrix} \phi_1 \\ \phi_2 \\ \phi_3 \\ \phi_4 \\ \phi_5 \\ \phi_6 \end{Bmatrix} = \begin{Bmatrix} 300 \\ 700 \\ 700 \\ 1200 \\ 1600 \\ 500 \end{Bmatrix}$$

The boundary conditions are $\phi_2 = 20$ and $\phi_4 = 50$.

Applying the second row of Eq. 7.14, the system equations are rewritten as

$$\begin{bmatrix} 44 & -12 & 3 & -5 & 0 & 0 \\ 0 & 58 & 0 & 0 & 0 & 0 \\ 3 & -4 & 32 & 7 & 3 & -1 \\ 0 & 0 & 0 & 48 & 0 & 0 \\ 0 & 6 & 3 & 2 & 60 & 20 \\ 0 & 0 & -1 & 11 & 20 & 62 \end{bmatrix} \begin{Bmatrix} \phi_1 \\ \phi_2 \\ \phi_3 \\ \phi_4 \\ \phi_5 \\ \phi_6 \end{Bmatrix} = \begin{Bmatrix} 300 \\ 1160 \\ 700 \\ 2400 \\ 1600 \\ 500 \end{Bmatrix}$$

where, for example, the second term in the force vector is found from

$$58 \times \phi_2 = 58 \times 20 = 1160$$

Now the columns of coefficients that multiply ϕ_2 and ϕ_4 are eliminated by transferring the terms to the right-hand side. For example, the first term in the force vector becomes

$$300 - (-12 \times 20) - (-5 \times 50) = 790$$

The final system of equations is calculated to be

$$\begin{bmatrix} 44 & 0 & 3 & 0 & 0 & 0 \\ 0 & 58 & 0 & 0 & 0 & 0 \\ 3 & 0 & 32 & 0 & 3 & -1 \\ 0 & 0 & 0 & 48 & 0 & 0 \\ 0 & 0 & 3 & 0 & 60 & 20 \\ 0 & 0 & -1 & 0 & 20 & 62 \end{bmatrix} \begin{Bmatrix} \phi_1 \\ \phi_2 \\ \phi_3 \\ \phi_4 \\ \phi_5 \\ \phi_6 \end{Bmatrix} = \begin{Bmatrix} 790 \\ 1160 \\ 430 \\ 2400 \\ 1380 \\ -50 \end{Bmatrix}$$

Note that the equations are still symmetric and banded, and are now ready to be solved.

7.5   Solution of the equations

When the boundary conditions have been incorporated into the system equations, the final step is the solution for the unknown variables. There are many techniques available, and these are discussed in detail in relevant mathematics and numerical analysis textbooks. Probably the most common methods for the equilibrium problems that have been discussed thus far are Gaussian elimination and Cholesky decomposition, which should be familiar to most students. However, there is one other novel solution technique that is widely used, particularly in commercial finite element packages, and that is the wavefront or frontal method. Since this method is so important and yet rarely described, it is discussed below.

Solution by the wavefront method

This technique does not strictly follow on from the previous two sections, because the system equations are never completely assembled when this method is used. The principles of a wavefront solution are as follows.

Firstly, the model is scanned to determine which element is first and which is last to use each of the nodes. A table of this information is stored for later use.

The element equations are then calculated in turn and assembled into a temporary matrix and vector. For example, the equations of element (2) might be

$$
\begin{bmatrix} a_{22} & a_{23} & a_{24} \\ a_{32} & a_{33} & a_{34} \\ a_{42} & a_{43} & a_{44} \end{bmatrix} \begin{Bmatrix} \phi_2 \\ \phi_3 \\ \phi_4 \end{Bmatrix} = \begin{Bmatrix} F_2 \\ F_3 \\ F_4 \end{Bmatrix}
\tag{7.15}
$$

which are added into the temporary matrix and vector as discussed in Section 7.3.

After the element's equations are added in, the nodes are checked for last appearances using the list derived above. When the last entry of a degree of freedom is noticed, the associated equation and corresponding column are removed by Gaussian elimination and written to a decomposed matrix file for later use. For example, assume the last occurrence of ϕ_3 is noted in the following equations:

$$
\begin{bmatrix} a_{11} & a_{12} & a_{13} & 0 & \dots \\ a_{21} & a_{22} & a_{23} & a_{24} \\ a_{31} & a_{32} & a_{33} & a_{34} \\ 0 & a_{42} & a_{43} & a_{44} \\ & \vdots \\ & \vdots \\ & \vdots \end{bmatrix} \begin{Bmatrix} \phi_1 \\ \phi_2 \\ \phi_3 \\ \phi_4 \\ \vdots \\ \vdots \\ \vdots \end{Bmatrix} = \begin{Bmatrix} F_1 \\ F_2 \\ F_3 \\ F_4 \\ \vdots \\ \vdots \\ \vdots \end{Bmatrix}
\tag{7.16}
$$

Since ϕ_3 does not occur in other elements, it is clear that

$$
\phi_3 = (F_3 - a_{31}\phi_1 - a_{32}\phi_2 - a_{34}\phi_4 - \dots)/a_{33}
\tag{7.17}
$$

This equation is stored for later evaluation, and also factored accordingly and subtracted from the other lines of the matrix in turn, resulting in the equations

$$
\begin{bmatrix} a'_{11} & a'_{12} & 0 & \dots \\ a'_{21} & a'_{22} & a'_{24} \\ 0 & a'_{42} & a'_{44} \\ & \vdots \\ & \vdots \\ & \vdots \end{bmatrix} \begin{Bmatrix} \phi_1 \\ \phi_2 \\ \phi_4 \\ \vdots \\ \vdots \\ \vdots \end{Bmatrix} = \begin{Bmatrix} F'_1 \\ F'_2 \\ F'_4 \\ \vdots \\ \vdots \\ \vdots \end{Bmatrix}
\tag{7.18}
$$

When the last occurrence of a specified variable (*i.e.* boundary condition) is detected, the associated equation is eliminated, and is rewritten to allow the calculation of the reaction. For example, if the last appearance of a specified displacement ϕ_2 is found with Eq. 7.18, then the reaction R_2 will be

$$R_2 = F'_2 - (a'_{21} \; a'_{22} \; a'_{24} \; ...) \begin{Bmatrix} \phi_1 \\ \phi_2 \\ \phi_4 \\ . \\ . \\ . \end{Bmatrix} \qquad [7.19]$$

The other terms in the matrix associated with ϕ_2 are then eliminated by adjusting the force vector terms. For example, the first force vector term becomes

$$F_1 - a_{12}\phi_2$$

When the last element has been considered, the last degree of freedom can be evaluated. Back substitution into the previously stored equations such as Eq. 7.17 reveals all the unknown degrees of freedom.

Finally, if the reactions are required, then the reaction equations (such as Eq. 7.19) are evaluated.

At any time there will only be a limited number of degrees of freedom in the temporary matrix. Consideration of the model shows that these degrees of freedom form a line across the model which gradually moves like a wave over the model; hence the name of the wavefront or frontal method. An example of the movement of a wavefront is presented in Section 3.8 and Fig. 3.23. To make the best use of the available computing power, the wavefront, and consequently the size of the temporary matrix, must be kept to a minimum. Clearly then the order in which the elements are considered is vital, and, to keep the size of the wavefront to a minimum, the elements must theoretically be labelled across the shortest dimension of a model. In practice, however, many commercial finite element programs include wavefront optimization routines. These allow the user either to manually specify the order in which the elements are assembled, or to let the computer automatically select the element order which (ideally) results in the minimum wavefront.

Problems

7.1 The beam element discussed in Example 7.1 has two degrees of freedom in its local coordinate system, namely two vertical translations and two rotations. If horizontal translations (u_i and u_j) are also included in the formulation, what transformation matrix would be required to use this new element in two-dimensional space?

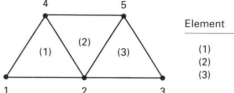

Element	Node		
	i	j	k
(1)	1	2	4
(2)	4	2	5
(3)	3	5	2

Figure 7.5

7.2 The stiffness matrices and force vectors of the three two-dimensional thermal elements given in Fig. 7.5 are as follows. Assemble the terms to produce the system equations.

$$[k^{(1)}] = [k^{(3)}] = \begin{bmatrix} 5 & 2 & 1 \\ 2 & 10 & 2 \\ 1 & 2 & 6 \end{bmatrix} \qquad [k^{(2)}] = \begin{bmatrix} 6 & 3 & 1 \\ 3 & 12 & 4 \\ 1 & 4 & 6 \end{bmatrix}$$

$$\{f^{(1)}\} = \{f^{(3)}\} = \begin{Bmatrix} 20 \\ 24 \\ 24 \end{Bmatrix} \qquad \{f^{(2)}\} = \begin{Bmatrix} 16 \\ 10 \\ 16 \end{Bmatrix}$$

7.3 The stiffness matrices and force vectors of the three plane truss elements given in Fig. 7.6 are as follows. Assemble the terms to produce the system equations.

$$[k^{(1)}] = \begin{bmatrix} 6 & 3 & -6 & -3 \\ 3 & 8 & -3 & -8 \\ -6 & -3 & 6 & 3 \\ -3 & -8 & 3 & 8 \end{bmatrix} \qquad \{f^{(1)}\} = \begin{Bmatrix} 3 \\ -4 \\ -3 \\ 4 \end{Bmatrix}$$

$$[k^{(2)}] = \begin{bmatrix} 2 & -5 & -2 & 5 \\ -5 & 3 & 5 & -3 \\ -2 & 5 & 2 & -5 \\ 5 & -3 & -5 & 3 \end{bmatrix} \qquad \{f^{(2)}\} = \begin{Bmatrix} 1 \\ 2 \\ -1 \\ -2 \end{Bmatrix}$$

$$[k^{(3)}] = \begin{bmatrix} 7 & 1 & -7 & -1 \\ 1 & 2 & -1 & -2 \\ -7 & -1 & 7 & 1 \\ -1 & -2 & 1 & 2 \end{bmatrix} \qquad \{f^{(3)}\} = \begin{Bmatrix} 5 \\ -4 \\ -5 \\ 4 \end{Bmatrix}$$

7.4 Adjust the system equations produced in Problem 7.2 for the boundary conditions $\phi_2 = 3$ and $\phi_5 = 0$.

7.5 Adjust the system equations produced in Problem 7.3 for the boundary conditions $u_2 - v_2 - 0$ and $v_3 = 0$.

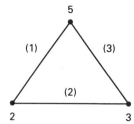

Element	Node	
	i	j
(1)	5	2
(2)	2	3
(3)	3	5

Figure 7.6

7.6 Prove that the solution of the system equations derived in Example 6.1 does indeed produce the stated answers.

7.7 Prove that the solution of the system equations derived in Example 6.7 does indeed produce the stated answers.

8   Higher-order element formulations

8.1   Introduction

The complete process of the finite element method is introduced in
Chapters 3 to 7 for the analysis of equilibrium problems in solid
mechanics and field problems. Throughout these chapters the work has
concentrated on simplex elements with linear interpolation functions, but
in practice higher-order elements are generally favoured because of their
increased accuracy. There is, however, a further step required with the
implementation of these higher-order formulations. Since explicit equations cannot be obtained for all the steps in the calculation of their
element equations, numerical integration may need to be performed.

This chapter introduces complex and multiplex higher-order elements.
Complex elements have the same general shape as simplex elements, but
contain extra nodes, usually on the mid-sides of the elements (Fig. 8.1).
Multiplex elements, however, are quadrilateral, hexahedron or wedge
shaped, as shown in Fig. 8.1.

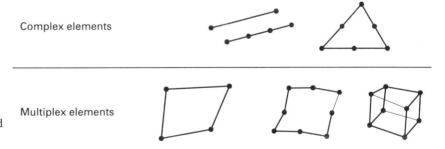

Complex elements

Multiplex elements

Figure 8.1 Typical complex and
multiplex elements

In addition to interpolation functions being used to describe how the
displacement or temperature varies within each element, they can also
define the geometry of the element. With simplex elements, linear
functions are used to describe both the variable distribution and the

geometry of the element. For example, with the two-dimensional triangular simplex element in Chapter 4, the displacements are defined by Eq. 4.32 as

$$\left\{ \begin{array}{c} u \\ v \end{array} \right\} = \left[\begin{array}{cccccc} N_i & 0 & N_j & 0 & N_k & 0 \\ 0 & N_i & 0 & N_j & 0 & N_k \end{array} \right] \left\{ \begin{array}{c} u_i \\ v_i \\ . \\ . \\ . \\ v_k \end{array} \right\}$$

while any position in the element can be specified by Eq. 4.24 as

$$\left\{ \begin{array}{c} x \\ y \end{array} \right\} = \left[\begin{array}{cccccc} L_1 & 0 & L_2 & 0 & L_3 & 0 \\ 0 & L_1 & 0 & L_2 & 0 & L_3 \end{array} \right] \left\{ \begin{array}{c} x_i \\ y_i \\ . \\ . \\ . \\ y_k \end{array} \right\}$$

Since it is shown that the local coordinates and the shape functions are equal for simplex elements (*i.e.* $L_1 = N_i$), then the two Eqs 4.24 and 4.32 are comparable.

This idea can be extended to complex and multiplex elements, where the order of the displacement interpolation function and the geometry interpolation function can be the same or different. Elements in which the functions are the same are known as isoparametric elements. If the order of the geometry is less than the field variable then the element is called subparametric, while if the reverse is true the element is superparametric. In practice, isoparametric elements are usually implemented, and the other element forms are rarely available.

The advantage of isoparametric formulations and their second-order (or higher) interpolation functions is that they not only represent the field variable more accurately, but also allow curved element boundaries (Fig. 8.2). This feature is particularly desirable, since few engineering components are composed solely of straight edges.

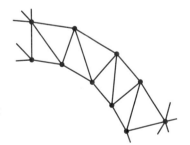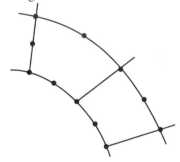

Figure 8.2 Modelling of curved boundary with simplex and complex elements

Finally, before higher-order elements are introduced, note that the node numbering scheme used to describe the specimen elements is changed in this chapter. In earlier chapters, letters are used to represent the nodes of different element types as they are introduced, for example i, j and k for the triangular element. This is done to emphasize the fact that the elements are general, and that each element in a model is just a repeat of the general element. For complex elements it is no longer practicable to label the nodes with letters because some solid brick elements can have up to 32 nodes. Consequently the specimen elements are now labelled numerically, but they should be considered in the same way as the simplex elements.

8.2   Natural coordinate systems and numerical integration

Whichever type of problem is being analysed by the finite element method, the calculations involve integrations of functions of the shape functions. For simplex and a few complex elements this is not a problem, because the shape functions are equal to or simple functions of the natural coordinates, and integration formulae are available that allow easy evaluation of functions of these coordinates. For many higher-order elements, however, this will not be possible and the integration will then be performed numerically.

The most suitable method for carrying out the integrations on the computer is Gauss quadrature, but the technique requires the limits of integration to be -1 to 1 or 0 to -1. Consequently a transformation must be applied to the standard governing equations (*i.e.* Eqs 5.22 and 5.23 or 6.45 and 6.46) before they can be evaluated. This is best achieved by the use of natural coordinates, which also allow the sides of the higher-order elements to be curved, and allow multiplex elements to have sides which are not parallel to the global coordinate system, as demonstrated later.

Figure 8.3 shows natural coordinate systems for triangular and quadrilateral elements. The area coordinates of the triangular elements have already been met in Chapter 4 when simplex elements were introduced. An area coordinate can have a value in the range 0 to 1. A different type of coordinate system is used for the quadrilateral element. The origin is defined at the intersection of the lines joining the mid-points of opposite sides, so that the coordinates range from -1 to 1.

It is important to note that the triangular element has curved sides in the global (x–y) coordinate system, but in the natural system the sides are straight. The same is true for the quadrilateral element, but also note that the sides of the element are parallel to the axes of the natural coordinate system, which is essential to ensure interelement continuity.

Therefore, natural coordinates allow distorted and curved elements to

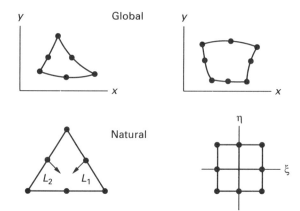

Figure 8.3 Appearance of isoparametric triangular and quadrilateral elements in global and natural coordinate systems

be included in finite element models, but they are also precisely what is required to implement the numerical integration by Gauss quadrature. The theory of Gauss quadrature is covered in most textbooks on numerical analysis, but is revised here because in many finite element packages the user can have some control over the process. The method works by evaluating the function to be integrated at a number of sampling points, multiplying these values by weighting functions and summing the products. The higher the order of the function, the more sampling points are required, but provided the correct number of points is used, the answer will be exact. To reduce the computer time in an analysis, some commercial packages allow the user to perform the integrations with a reduced number of sampling points. This leads to an approximate answer, although in many cases it is acceptable.

Numerical integration

When a function is integrated over a given range, the process is simply calculating the area under the graph formed by the function as illustrated in Fig. 8.4. Like other numerical techniques such as the trapezoidal method and Simpson's method, Gauss quadrature calculates the value of the function at a point and multiplies by a length along the ξ axis to evaluate the area calculations. Hence, if the function is a linear function of ξ (*i.e.* a straight line), the value of the function at $\xi=0$ multiplied by 2, the length along the ξ axis, would give the area precisely. So one sampling point is required if the function is of the first order. In general n sampling points will evaluate a polynomial of order $2n-1$ exactly.

Gauss quadrature is expressed as

$$\int_{-1}^{1} f(\xi)\,\mathrm{d}\xi = \sum_{i=1}^{n} H_i f(\xi_i) \qquad [8.1]$$

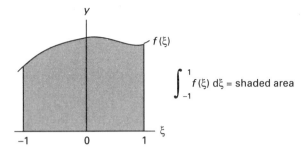

$$\int_{-1}^{1} f(\xi)\, d\xi = \text{shaded area}$$

Figure 8.4 Integration of the function $f(\xi)$

where n is the number of integration points and H_i is the weighting function at sampling point i.

The location of the sampling points and the value of the weighting functions for Gaussian integration with up to four points are shown in Table 8.1 with limits -1 to 1 and 0 to 1. Note that the data are given to a high degree of accuracy, and indeed even more precise values are available in the literature.

Table 8.1. Gauss quadrature information: $\int f(\xi)d\xi = \Sigma_{i=1}^{n} H_i f(\xi_i)$ for two ranges of integral limits

Degree of polynomial	Number of points n	Sampling point ξ_i	Weighting function H_i
Limits −1 to 1			
1	1	0.0	2.0
≤3	2	±0.577 350	1.0
≤5	3	0.0	0.888 889
		±0.774 597	0.555 556
≤7	4	±0.861 136	0.347 855
		±0.339 981	0.652 145
Limits 0 to 1			
1	1	0.5	1.0
≤3	2	0.211 325	0.5
		0.788 675	0.5
≤5	3	0.112 702	0.277 778
		0.5	0.444 444
		0.887 298	0.277 778
≤7	4	0.069 432	0.173 928
		0.330 009	0.326 072
		0.669 990	0.326 072
		0.930 568	0.173 928

Example 8.1: numerical integration

Evaluate the following integral by long-hand and by the Gauss quadrature method:

$$I = \int_{-1}^{1} (3x^5 + 4x^4 + 9x^2 + 15)\, dx$$

Integrating in the usual way gives

$$I = \left[\frac{3x^6}{6} + \frac{4x^5}{5} + \frac{9x^3}{3} + 15x \right]_{-1}^{1} = \left[19.3 - (-18.3) \right] = 37.6$$

Using the Gauss method, three sampling points are required:

Position, x	$f(x)$	H_x	$H_x f(x)$
$-0.774\ 597$	$21.003\ 441$	$0.555\ 556$	$11.668\ 588$
0.0	$15.000\ 000$	$0.888\ 889$	$13.333\ 335$
$0.774\ 597$	$22.676\ 573$	$0.555\ 556$	$12.598\ 106$

$$I = 37.600\ 029$$

The two results are clearly very close. Use of more precise sampling data and more accurate calculations would reduce the error still further.

Note that for this example, if a reduced integration scheme of two sampling points is used, the answer becomes 36.888 882, which has an error of less than 2 per cent. The 30 per cent reduction in the number of calculations that this answer requires might easily make this degree of inaccuracy acceptable.

8.3   Higher-order one-dimensional elements

This section introduces one-dimensional complex elements and some of the techniques required to deal with higher-order formulations. In practice such one-dimensional elements are rarely used, but they are a very useful means of introducing the concepts of complex elements.

Two types of natural coordinates are possible with these one-dimensional elements, allowing the element equations to be derived in two ways. In the sections below, calculations using the natural coordinate system developed in Chapter 4 are considered initially and in some detail. Following that, the derivation using a second coordinate system is described briefly, since it leads on to the methods required for two- and three-dimensional multiplex elements.

8.3.1   Quadratic and cubic elements

As discussed in Section 8.2, to allow the sides of complex and multiplex elements to be curved, the elements must be specified and developed in natural coordinates, with some type of transformation then used to apply the elements in the global system. Consequently, the shape functions must be devised in the same coordinates. A one-dimensional element with three nodes and a quadratic interpolation function is shown in Fig. 8.5, and includes the natural coordinate system first introduced in Section 4.4. Although two coordinates are shown they are not independent, since clearly

$$L_1 + L_2 = 1 \qquad [8.2]$$

The interpolation function in terms of the coordinate L_2 (which ranges from 0 to 1) will be

$$\phi = a_1 + a_2 L_2 + a_3 L_2^2 \qquad [8.3]$$

where a_1, a_2 and a_3 are constants as usual, and are determined by substituting in the nodal values of the field variable,

$$\phi = \phi_1 \quad \text{at} \quad L_2 = 0$$
$$\phi = \phi_2 \quad \text{at} \quad L_2 = 0.5$$
$$\phi = \phi_3 \quad \text{at} \quad L_2 = 1$$

The three simultaneous equations that result prove that

$$a_1 = \phi_1$$
$$a_2 = (-3\phi_1 + 4\phi_2 - \phi_3)/L \qquad [8.4]$$
$$a_3 = (2\phi_1 - 4\phi_2 + 2\phi_3)/L^2$$

When these are substituted back into Eq. 8.3 and rearranged, one obtains the standard form of the interpolation function in terms of the shape functions:

$$\phi = N_1\phi_1 + N_2\phi_2 + N_3\phi_3 \qquad [8.5]$$

where

$$N_1 = 1 - 3L_2 + 2L_2^2 = L_1(2L_1 - 1)$$
$$N_2 = 4L_2(1 - L_2) = 4L_1L_2 \qquad [8.6]$$
$$N_3 = L_2(2L_2 - 1)$$

The shape functions are quadratic functions of the coordinate system, and in common with all other shape functions they equal unity at their associated nodes and zero at the others.

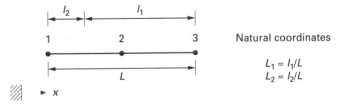

Figure 8.5 One-dimensional element with quadratic interpolation function

If the same procedures are carried out for a one-dimensional element with a cubic interpolation function (and four nodes), the following relationships are derived:

$$\phi = N_1\phi_1 + N_2\phi_2 + N_3\phi_3 + N_4\phi_4 \qquad [8.7]$$

where

$$
\begin{aligned}
N_1 &= L_1(1 - 4.5L_1L_2) \\
N_2 &= 4.5L_1L_2(3L_1 - 1) \\
N_3 &= 4.5L_1L_2(3L_2 - 1) \\
N_4 &= L_2(1 - 4.5L_1L_2)
\end{aligned}
\qquad [8.8]
$$

The variation of N_1 and N_2 down the length of the element is shown in Fig. 8.6. The curves are cubic functions of the coordinates.

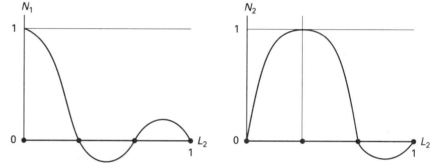

Figure 8.6 Variation of N_1 and N_2 in a cubic element

8.3.2 Evaluation of the element equations

The general finite element equations for stress analysis and field problems are derived in Chapters 5 and 6 respectively. The calculation of the element equations involves the evaluation of integrals such as

$$\int_V [B]^T[D][B]\mathrm{d}V \qquad \int_S h[N]^T[N]\mathrm{d}S \qquad \int_S q[N]^T\mathrm{d}S$$

The $[B]$ matrix is calculated by the proper differentiation of the shape functions. For example, for a one-dimensional thermal problem using a quadratic element,

$$[B] = \left[\ \frac{\mathrm{d}N_1}{\mathrm{d}x}\ \frac{\mathrm{d}N_2}{\mathrm{d}x}\ \frac{\mathrm{d}N_3}{\mathrm{d}x}\ \right]$$

With simplex elements this step is straightforward, but with complex elements the shape functions are functions of the (local) natural coordinate system, and not the global $(x-y)$ system. Clearly then some form of transformation is required. It is known that the geometry of the cubic element is defined by the equation

$$x = N_1 x_1 + N_2 x_2 + N_3 x_3 \qquad [8.9]$$

where x_1, x_2 and x_3 are the global coordinates of the three nodes.

Therefore, by the chain rule of differentiation, since

$$\frac{dN_1}{dL_2} = \frac{dN_1}{dx} \frac{dx}{dL_2} \qquad [8.10]$$

the derivative of the shape function N_1 with respect to the global x must be

$$\frac{dN_1}{dx} = \left(1 \Big/ \frac{dx}{dL_2} \right) \frac{dN_1}{dL_2} \qquad [8.11]$$

The terms on the right-hand side of Eq. 8.11 are obtained directly from Eqs 8.6 and 8.9. The quantity dx/dL_2 is called the Jacobian matrix for the transformation equation, and is usually denoted by $[J]$. For the one-dimensional element it is a $[1 \times 1]$ matrix; for two and three dimensions it proves to be $[2 \times 2]$ and $[3 \times 3]$ respectively. Hence Eq. 8.11 can be written as

$$\frac{dN_1}{dx} = [J]^{-1} \frac{dN_1}{dL_2} \qquad [8.12]$$

where

$$[J] = \frac{dx}{dL_2} = \frac{dN_1}{dL_2} x_1 + \frac{dN_2}{dL_2} x_2 + \frac{dN_3}{dL_2} x_3 \qquad [8.13]$$

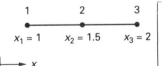

1   2   3

$x_1 = 1$ $x_2 = 1.5$ $x_3 = 2$

x

Figure 8.7

Example 8.2: calculation of a $[B]$ matrix

Calculate the $[B]$ matrix for the one-dimensional quadratic element shown in Fig. 8.7.

The shape functions for this element are derived in the previous section. Equation 8.6 shows that

$$\begin{aligned} N_1 &= 1 - 3L_2 + 2L_2^2 \\ N_2 &= 4L_2(1 - L_2) \\ N_3 &= L_2(2L_2 - 1) \end{aligned} \qquad [8.14]$$

Therefore the Jacobian matrix as defined by Eq. 8.13 is calculated by

$$\frac{dN_1}{dL_2} = -3 + 4L_2$$

$$\frac{dN_2}{dL_2} = 4 - 8L_2 \qquad\qquad [8.15]$$

$$\frac{dN_3}{dL_2} = 4L_2 - 1$$

Hence

$$[J] = (-3 + 4L_2) \times 1 + (4 - 8L_2) \times 1.5 + (4L_2 - 1) \times 2 = 1$$
and $[J]^{-1} = 1$

Using Eq. 8.12, the $[B]$ matrix is then calculated as

$$[B] = \left[\frac{dN_1}{dx} \; \frac{dN_2}{dx} \; \frac{dN_3}{dx} \right]$$

$$= 1[4L_2 - 3 \; 4 - 8L_2 \; 4L_2 - 1] \qquad\qquad [8.16]$$

Note that the Jacobian is a constant for this simple element. This is not the case in two- and three-dimensional analyses where $[J]$ will usually be a function of the local coordinates, and an explicit inverse cannot be obtained. This is the reason why numerical integration must be employed.

In fact, the Jacobian is a constant only if node 2 of the element is precisely at the mid-side of the element. For example, if $x_2 = 1.4$ rather than 1.5, then

$$[J] = 0.6 + 0.8L_2$$

The calculation of the $[B]$ matrix is still straightforward for this one-dimensional element, but the calculation of the stiffness matrix is no longer trivial. (Also note the special case of $x = 1.75$, giving $[J] = 2 - 2L_2$, which equals zero at $L_2 = 1$.)

Further discussion on moving the mid-side nodes of elements is included in Section 9.4, which deals with element distortion and its effect on the accuracy of the solution.

Before the element equations can be evaluated, one more transformation is required. The volume and surface integrals of the general element equations must be expressed in terms of natural coordinates with appropriate changes in the limits of integration. This is achieved by the following type of substitution:

$$\int_L f(x)dx = \int_0^1 g(L_2)|J|dL_2 \qquad\qquad [8.17]$$

where $g(L_2)$ is the function $f(x)$ written in terms of the natural coordinate system L_2, and $|J|$ is the determinant of the Jacobian. This change of variable holds providing the Jacobian's determinant $|J|$ is greater than zero.

Example 8.3: calculation of a [k] matrix
Calculate the stiffness matrix for the one-dimensional thermal element in Example 8.2.

The stiffness matrix can be written as

$$[k] = \int_V [B]^T[D][B]dV = A \int_0^1 [B]^T[D][B]|J|dL_2 \qquad [8.18]$$

(assuming that the cross-sectional area is constant). The $[B]$ matrix was calculated in Example 8.2, and the $[D]$ matrix will be equal to the thermal conductivity K. Also, since the Jacobian was found to equal 1, Eq. 8.18 becomes

$$[k] = AK \int_0^1 \begin{Bmatrix} 4L_2 - 3 \\ 4 - 8L_2 \\ 4L_2 - 1 \end{Bmatrix} [4L_2 - 3 \;\; 4 - 8L_2 \;\; 4L_2 - 1]dL_2 \qquad [8.19]$$

$$= AK \begin{bmatrix} 2.33 & -2.67 & 0.33 \\ -2.67 & 5.33 & -2.67 \\ 0.33 & -2.67 & 2.33 \end{bmatrix} \qquad [8.20]$$

For this one-dimensional example the integration can be performed exactly, because the Jacobian is a constant.

If the $[B]$ matrix and the Jacobian are not such simple functions (*i.e.* the internal node is not at the centre of the element), then numerical integration is required. To illustrate the method, Eq. 8.19 is now recalculated using numerical integration.

Two sampling points are required since the expanded terms are second order. The locations of the sampling points are then

$$L_2 = 0.211\ 325 \quad \text{and} \quad L_2 = 0.788\ 675$$

with weighting functions of 0.5 in both cases (from Table 8.1). Hence the stiffness matrix is calculated as follows:

$$[k] = AK \begin{bmatrix} 0.5 \times \begin{Bmatrix} -2.1547 \\ 2.3094 \\ -0.1547 \end{Bmatrix} [-2.1547 \;\; 2.3094 \;\; -0.1547] \\ \\ + 0.5 \times \begin{Bmatrix} 0.1547 \\ -2.3094 \\ 2.1547 \end{Bmatrix} [0.1547 \;\; -2.3094 \;\; 2.1547] \end{bmatrix}$$

$$= \frac{AK}{2} \left[\begin{bmatrix} 4.6427 & -4.9760 & 0.3333 \\ -4.9760 & 5.3333 & -0.3572 \\ 0.3333 & -0.3572 & 0.0239 \end{bmatrix} \right.$$

$$\left. + \begin{bmatrix} 0.0239 & -0.3572 & 0.3333 \\ -0.3572 & 5.3333 & -4.9760 \\ 0.3333 & -4.9760 & 4.6427 \end{bmatrix} \right]$$

$$= AK \begin{bmatrix} 2.33 & -2.67 & 0.33 \\ -2.67 & 5.33 & -2.67 \\ 0.33 & -2.67 & 2.33 \end{bmatrix}$$

which is exactly the same as Eq. 8.20.

The evaluation of terms such as

$$\int_S h[N]^{\mathrm{T}}[N] \mathrm{d}S \quad \text{and} \quad \int_S q[N]^{\mathrm{T}} \mathrm{d}S$$

is more straightforward than the terms involving the $[B]$ matrix because there is no transformation required. The shape functions are functions of the local coordinates, and fortunately the integration formulae introduced in Chapter 4 can still be applied to these integrals. This is illustrated in the following example.

Example 8.4: calculation of a perimeter convention term
What term must be included in the stiffness matrix of a quadratic thermal element to take account of convection from the perimeter?

The basic term required (from Eq. 6.45) is

$$\int_S h[N]^{\mathrm{T}}[N] \mathrm{d}S \tag{8.21}$$

The shape function matrix (from Eq. 8.6) is

$$\begin{aligned} [N] &= [N_1 \ N_2 \ N_3] \\ &= [L_1(2L_1 - 1) \ 4L_1L_2 \ L_2(2L_2 - 1)] \end{aligned} \tag{8.22}$$

so that Eq. 8.21 becomes

$$\int_L h \begin{bmatrix} N_1^2 & N_1N_2 & N_1N_3 \\ N_2N_1 & N_2^2 & N_2N_3 \\ N_3N_1 & N_3N_2 & N_3^2 \end{bmatrix} P \, \mathrm{d}x \tag{8.23}$$

where $\mathrm{d}S = P \, \mathrm{d}x$ is substituted for the surface area, if the perimeter length P does not vary down the length of the element. Using the definition of the shape functions in Eq. 8.22, the integral can then be evaluated term by term with the integration formula of Eq. 4.20, namely

$$\int_L L_1^\alpha L_2^\beta dx = \frac{\alpha!\beta!}{(\alpha + \beta + 1)!} L$$

For the first coefficient in Eq. 8.23,

$$\int_L hP(L_1(2L_1 - 1))^2 dx = \int_L hP(4L_1^4 - 4L_1^3 + L_1^2) dx$$

$$= hP\left(4 \times \frac{L}{5} - 4 \times \frac{L}{4} + \frac{L}{3}\right) = \frac{2hPL}{15}$$

If the other coefficients are dealt with in the same way, the complete term is found to be

$$\frac{hPL}{30}\begin{bmatrix} 4 & 2 & -1 \\ 2 & 16 & 2 \\ -1 & 2 & 4 \end{bmatrix} \qquad [8.24]$$

Thus natural coordinates and numerical integration are required for complex elements. Clearly the number of calculations required is significantly increased compared with simplex elements, but the accuracy is considerably improved. Once the element equations have been calculated, the analysis proceeds in the same way as before, as the following example demonstrates.

Example 8.5: straight fin analysis

Reanalyse the problem of the straight fin in Example 6.1 (shown again in Fig. 8.8) using quadratic elements.

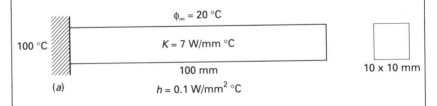

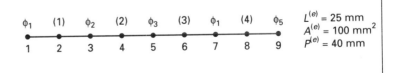

Figure 8.8 (a) One-dimensional fin (b) finite element idealization with four quadratic elements

The following element equations may be used:

$$[k] = \frac{AK}{3L} \begin{bmatrix} 7 & -8 & 1 \\ -8 & 16 & -8 \\ 1 & -8 & 7 \end{bmatrix} + \frac{hPL}{30} \begin{bmatrix} 4 & 2 & -1 \\ 2 & 16 & 2 \\ -1 & 2 & 4 \end{bmatrix}$$

$$+ hA \begin{bmatrix} 0 & 0 & 0 \\ 0 & 0 & 0 \\ 0 & 0 & 1 \end{bmatrix} \tag{8.25}$$

$$\{F\} = \frac{h\phi_\infty PL}{6} \begin{Bmatrix} 1 \\ 4 \\ 1 \end{Bmatrix} + h\phi_\infty A \begin{Bmatrix} 0 \\ 0 \\ 1 \end{Bmatrix} \tag{8.26}$$

where all the constants have their usual meanings, and the terms in the stiffness matrix are from conduction, perimeter convection and end convection respectively. Two of these terms were derived in Examples 8.3 and 8.4. The terms in the force vector are due to perimeter convection and end convection.

For element (1),

$$[k^{(1)}] = 9.334 \begin{bmatrix} 7 & -8 & 1 \\ -8 & 16 & -8 \\ 1 & -8 & 7 \end{bmatrix} + 3.334 \begin{bmatrix} 4 & 2 & -1 \\ 2 & 16 & 2 \\ -1 & 2 & 4 \end{bmatrix}$$

$$= \begin{bmatrix} 78.674 & -68.004 & 6.000 \\ -68.004 & 202.688 & -68.004 \\ 6.000 & -68.004 & 78.674 \end{bmatrix}$$

$$\{F^{(1)}\} = 333.333 \begin{Bmatrix} 1 \\ 4 \\ 1 \end{Bmatrix}$$

For elements (2) and (3) the same equations apply. However, element (4) can also lose heat by convection through node 9; consequently the last term in Eqs 8.25 and 8.26 must also be included in its equations, so that

$$[k^{(4)}] = \begin{bmatrix} 78.674 & -68.004 & 6.000 \\ -68.004 & 202.688 & -68.004 \\ 6.000 & -68.004 & 88.674 \end{bmatrix}$$

$$\{F^{(4)}\} = \begin{Bmatrix} 333.333 \\ 1333.333 \\ 533.333 \end{Bmatrix}$$

When these equations are combined in the usual way to give the system equations, and solved with the boundary condition $\phi_1 = 100\,°\text{C}$, the nodal temperatures are found to be

$$\{\Phi_S\}^T = [100.00 \;\; 50.97 \;\; 32.32 \;\; 24.77 \;\; 21.89 \;\; 20.73 \;\; 20.29$$
$$20.12 \;\; 20.07]$$

The heat applied from the surroundings at node 1 to maintain the constant temperature of 100 °C is found to be 4261.57 watts.

The following comparison of these results with those calculated in Example 6.1 shows the advantage of the quadratic elements. The model developed in this section with complex elements leads to significantly more accurate results than the model using simplex elements. Indeed if the problem is modelled with only two quadratic elements, the results are comparable with the model using four simplex elements.

Temperatures (°C)	ϕ_1	ϕ_3	ϕ_5	ϕ_7	ϕ_9
4 simplex elements (Example 6.1)	100.00	27.45	20.69	20.06	20.01
4 quadratic elements (Example 8.4)	100.00	32.32	21.90	20.29	20.07
2 quadratic elements	100.00	31.08	23.70	20.53	20.29
Theory	100.00	32.08	21.83	20.28	20.07
x location (mm)	0	25.0	50.0	75.0	100.0

8.3.3   An alternative formulation

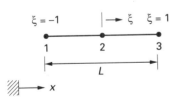

Figure 8.9 Alternative one-dimensional coordinate system

An alternative one-dimensional coordinate system to that previously used is shown in Fig. 8.9; the coordinate ξ varies from -1 to 1. The interpolation function for the element in terms of ξ will be

$$\phi = \alpha_1 + \alpha_2\xi + \alpha_3\xi^2 \qquad [8.27]$$

If the α constants are derived in the usual way by substituting in the nodal values of the variable, the interpolation function can be written as

$$\phi = N_1\phi_1 + N_2\phi_2 + N_3\phi_3$$

where

$$N_1 = \frac{\xi}{2}(\xi - 1)$$
$$N_2 = -(\xi + 1)(\xi - 1) \qquad [8.28]$$
$$N_3 = \frac{\xi}{2}(\xi + 1)$$

These shape functions display the features expected of all shape functions, and are quadratic functions of the coordinate system since the interpolation function is a quadratic.

The derivation of the element equations using this coordinate system then proceeds in a similar way to that discussed in the previous sections, except that the use of local integration formulae is no longer possible. The [B] matrix for such an element is defined as

$$[B] = \left[\frac{dN_1}{dx} \frac{dN_2}{dx} \frac{dN_3}{dx} \right] \qquad [8.29]$$

However, since the shape functions are defined in terms of ξ and not x, a transformation is required between the local and global coordinate systems. The relationship linking the two systems is found from

$$x = N_1 x_1 + N_2 x_2 + N_3 x_3 \qquad [8.30]$$

where x_1, x_2 and x_3 are the (global) coordinates of the three nodes, and therefore the derivatives of the shape functions are calculated as follows:

$$\frac{dN_\beta}{dx} = \frac{dN_\beta}{d\xi} \frac{d\xi}{dx} = \left(1/\frac{dx}{d\xi} \right) \frac{dN_\beta}{d\xi} \qquad [8.31]$$

This equation is comparable to Eq. 8.11 and the quantity $dx/d\xi$ is of course the Jacobian, which is easily evaluated through Eqs 8.28 and 8.30. The Jacobian is also required to change the variable of integration in the calculation of the element equations:

$$\int_L f(x)dx = \int_{-1}^{1} g(\xi)|J|d\xi \qquad [8.32]$$

In fact, providing the mid-side node is precisely half-way along the length of the element, the Jacobian proves to be the length of the element divided by two.

Example 8.6: calculation of a [k] matrix using an alternative coordinate system

Evaluate the stiffness matrix for the one-dimensional quadratic element considered in Examples 8.2 and 8.3.

Firstly, the Jacobian must be calculated:

$$[J] = \frac{\mathrm{d}x}{\mathrm{d}\xi} = \frac{\mathrm{d}N_1}{\mathrm{d}\xi}x_1 + \frac{\mathrm{d}N_2}{\mathrm{d}\xi}x_2 + \frac{\mathrm{d}N_3}{\mathrm{d}\xi}x_2$$

$$= (\xi - \tfrac{1}{2}) \times 1 + (-2\xi) \times 1.5 + (\xi + \tfrac{1}{2}) \times 2 = \tfrac{1}{2}$$

$$[J]^{-1} = 2$$

Therefore using Eqs 8.29 and 8.31,

$$[B] = 2 \left[\xi - \tfrac{1}{2} \quad -2\xi \quad \xi + \tfrac{1}{2} \right] = [2\xi - 1 \quad -4\xi \quad 2\xi + 1] \qquad [8.33]$$

Hence the stiffness matrix is calculated as

$$[k] = \int_V [B]^{\mathrm{T}}[D][B]\mathrm{d}V = A \int_{-1}^{1} [B]^{\mathrm{T}}[D][B]\mathrm{d}x \, |J|\mathrm{d}\xi$$

$$= \frac{AK}{2} \int_{-1}^{1} \left\{ \begin{array}{c} 2\xi - 1 \\ -4\xi \\ 2\xi + 1 \end{array} \right\} [2\xi - 1 \quad -4\xi \quad 2\xi + 1]\mathrm{d}\xi \qquad [8.34]$$

$$= AK \left[\begin{array}{ccc} 2.33 & -2.67 & 0.33 \\ -2.67 & 5.33 & -2.67 \\ 0.33 & -2.67 & 2.33 \end{array} \right]$$

where the variable of integration is changed by the substitution in Eq. 8.32.

The stiffness matrix is the same as that calculated in Examples 8.2 and 8.3. As with the previous derivation, numerical integration is not required for the simple element. However, for distorted elements of this type, and more complex two- and three-dimensional elements, numerical integration will be needed.

8.4   Higher-order two- and three-dimensional elements

8.4.1 Isoparametric triangular elements

The natural coordinates of two-dimensional triangular elements are area coordinates (see Section 4.4), and these must be used in the definition of

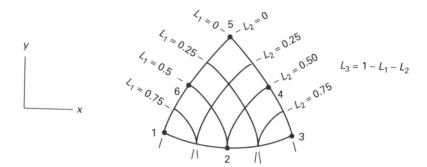

Figure 8.10 Natural coordinates of the triangular element

the shape functions. There are two independent coordinates, L_1 and L_2, as shown in Fig. 8.10.

Shape functions

The interpolation function required for the element in Fig. 8.10 is

$$\phi = N_1\phi_1 + N_2\phi_2 + N_3\phi_3 + N_4\phi_4 + N_5\phi_5 + N_6\phi_6 \qquad [8.35]$$

where the shape functions are found to be

$$
\begin{aligned}
N_1 &= L_1(2L_1 - 1) \\
N_2 &= 4L_1L_2 \\
N_3 &= L_2(2L_2 - 1) \\
N_4 &= 4L_2L_3 = 4L_2(1 - L_1 - L_2) \\
N_5 &= L_3(2L_3 - 1) = 1 - 3(L_1 + L_2) + 2(L_1 + L_2)^2 \\
N_6 &= 4L_3L_1 = 4L_1(1 - L_1 - L_2)
\end{aligned}
\qquad [8.36]
$$

In common with all shape functions they equal one at their associated node and zero at the others, as illustrated in Fig. 8.11.

Since the element is isoparametric, the geometry of the element can be expressed in a similar way to the field variable:

$$
\left\{ \begin{array}{c} x \\ y \end{array} \right\} =
\left[\begin{array}{cccccc}
N_1 & 0 & N_2 & 0 & \dots & N_6 & 0 \\
0 & N_1 & 0 & N_2 & & 0 & N_6
\end{array} \right]
\left\{ \begin{array}{c} x_1 \\ y_1 \\ x_2 \\ \cdot \\ \cdot \\ \cdot \\ y_6 \end{array} \right\}
\qquad [8.37]
$$

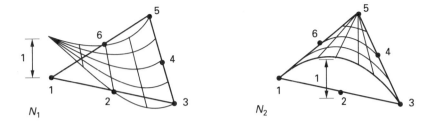

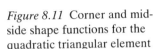

Figure 8.11 Corner and mid-side shape functions for the quadratic triangular element

Coordinate transformation

To calculate the [B] matrix, the partial derivatives of the shape functions are required. For example, for a thermal element with the interpolation function defined in Eq. 8.35, the [B] matrix is found from

$$\{g\} = \begin{Bmatrix} \dfrac{\partial \phi}{\partial x} \\[2mm] \dfrac{\partial \phi}{\partial y} \end{Bmatrix} = \begin{bmatrix} \dfrac{\partial N_1}{\partial x} & \dfrac{\partial N_2}{\partial x} & \dfrac{\partial N_3}{\partial x} & \dfrac{\partial N_4}{\partial x} & \dfrac{\partial N_5}{\partial x} & \dfrac{\partial N_6}{\partial x} \\[2mm] \dfrac{\partial N_1}{\partial y} & \dfrac{\partial N_2}{\partial y} & \dfrac{\partial N_3}{\partial y} & \dfrac{\partial N_4}{\partial y} & \dfrac{\partial N_5}{\partial y} & \dfrac{\partial N_6}{\partial y} \end{bmatrix} \begin{Bmatrix} \phi_1 \\ \phi_2 \\ \phi_3 \\ \phi_4 \\ \phi_5 \\ \phi_6 \end{Bmatrix}$$

$$= [B]\{\Phi\} \qquad\qquad [8.38]$$

However, since the shape functions are defined in terms of L_1 and L_2, a transformation is required. As with the one-dimensional element, the chain law for differentiation must be used. Consider for example the first shape function N_1:

$$\begin{aligned} \frac{\partial N_1}{\partial L_1} &= \frac{\partial N_1}{\partial x}\frac{\partial x}{\partial L_1} + \frac{\partial N_1}{\partial y}\frac{\partial y}{\partial L_1} \\[2mm] \frac{\partial N_1}{\partial L_2} &= \frac{\partial N_1}{\partial x}\frac{\partial x}{\partial L_2} + \frac{\partial N_1}{\partial y}\frac{\partial y}{\partial L_2} \end{aligned} \qquad\qquad [8.39]$$

This can be rewritten in the form

$$\begin{Bmatrix} \dfrac{\partial N_1}{\partial x} \\[2mm] \dfrac{\partial N_1}{\partial y} \end{Bmatrix} = [J]^{-1} \begin{Bmatrix} \dfrac{\partial N_1}{\partial L_1} \\[2mm] \dfrac{\partial N_1}{\partial L_2} \end{Bmatrix} \qquad\qquad [8.40]$$

where the Jacobian $[J]$ is

$$[J] = \begin{bmatrix} \dfrac{\partial x}{\partial L_1} & \dfrac{\partial y}{\partial L_1} \\ \dfrac{\partial x}{\partial L_2} & \dfrac{\partial y}{\partial L_2} \end{bmatrix}$$ [8.41]

The other terms of Eq. 8.38 can be dealt with in the same way to give a similar relationship to that in Eq. 8.40.

Consideration of Eqs 8.36 and 8.37 reveals that the Jacobian will be a function of L_1 and L_2, and an explicit inverse will not be obtainable, as illustrated in Example 8.7.

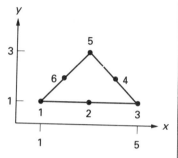

Figure 8.12

Example 8.7: the Jacobian of a two-dimensional element
Derive the general form of the Jacobian for a perfectly shaped quadratic triangular element. If the coordinates of the corner nodes of such an element are (1,1), (5,1) and (3,3), as illustrated in Fig. 8.12, then evaluate the Jacobian and its determinant at $L_1 = L_2 = 0.5$.

According to Eqs 8.37 and 8.41,

$$[J] = \begin{bmatrix} \dfrac{\partial x}{\partial L_1} & \dfrac{\partial y}{\partial L_1} \\ \dfrac{\partial x}{\partial L_2} & \dfrac{\partial y}{\partial L_2} \end{bmatrix} = \begin{bmatrix} \dfrac{\partial N_1}{\partial L_1} & \dfrac{\partial N_2}{\partial L_1} & \dfrac{\partial N_3}{\partial L_1} & \cdots & \dfrac{\partial N_6}{\partial L_1} \\ \dfrac{\partial N_1}{\partial L_2} & \dfrac{\partial N_2}{\partial L_2} & \dfrac{\partial N_3}{\partial L_2} & \cdots & \dfrac{\partial N_6}{\partial L_2} \end{bmatrix} \begin{bmatrix} x_1 \ y_1 \\ x_2 \ y_2 \\ x_2 \ y_3 \\ \cdot \ \cdot \\ \cdot \ \cdot \\ \cdot \ \cdot \\ x_6 \ y_6 \end{bmatrix}$$ [8.42]

The shape functions are defined in Eq. 8.36 and are readily differentiated with respect to the natural coordinates L_1 and L_2:

$$\frac{\partial N_1}{\partial L_1} = 4L_1 - 1 \qquad\qquad \frac{\partial N_2}{\partial L_1} = 4L_2$$

$$\frac{\partial N_3}{\partial L_1} = 0 \qquad\qquad \frac{\partial N_4}{\partial L_1} = -4L_2$$

$$\frac{\partial N_5}{\partial L_1} = -3 + 4(L_1 + L_2) \qquad \frac{\partial N_6}{\partial L_1} = 4 - 8L_1 - 4L_2$$

$$\frac{\partial N_1}{\partial L_2} = 0 \qquad\qquad \frac{\partial N_2}{\partial L_2} = 4L_1$$

$$\frac{\partial N_3}{\partial L_2} = 4L_2 - 1 \qquad\qquad \frac{\partial N_4}{\partial L_2} = 4 - 8L_2 - 4L_1$$

$$\frac{\partial N_5}{\partial L_2} = -3 + 4(L_1 + L_2) \qquad \frac{\partial N_6}{\partial L_2} = -4L_1 \qquad\qquad [8.43]$$

Substitution of these terms into Eq. 8.42 then gives the general form of the Jacobian. If the side nodes are exactly at the mid-side position then the Jacobian proves to be a constant, but if the element sides are distorted then the Jacobian is a function of the coordinates L_1 and L_2, without an explicit inverse.

Consider the element specified, which is shown in Fig. 8.12. At the specified point $L_1 = L_2 = 0.5$, Eq. 8.42 becomes

$$[J] = \begin{bmatrix} 1 & 2 & 0 & -2 & 1 & -2 \\ 0 & 2 & 1 & -2 & 1 & -2 \end{bmatrix} \begin{bmatrix} 1 & 1 \\ 3 & 1 \\ 5 & 1 \\ 4 & 2 \\ 3 & 3 \\ 2 & 2 \end{bmatrix} = \begin{bmatrix} -2 & -2 \\ 2 & -2 \end{bmatrix}$$

Consequently the Jacobian's determinant is

$$|J| = (-2 \times -2) - (-2 \times 2) = 8$$

Changing the variable of integration

Since the shape functions are defined in terms of the natural coordinates, the variables of integration in the finite element equations derived in Chapters 5 and 6 need to be changed from the global Cartesian system to the natural one. This is straightforward, and involves the Jacobian matrix as in the one-dimensional elements,

$$\int_A f(x,y)\mathrm{d}x\ \mathrm{d}y = \int_0^1\!\!\int_0^1 g(L_1,L_2)|J|\mathrm{d}L_1\ \mathrm{d}L_2 \qquad\qquad [8.44]$$

where $g(L_1,L_2)$ is the function $f(x,y)$ written in terms of the natural coordinates L_1 and L_2. Since the Jacobian might be a function of L_1 and L_2, this integration is normally performed numerically.

Numerical integration

When Gauss quadrature is applied over a triangular region, the sampling points and weighting functions given in Table 8.2 should be used.

Table 8.2. Gauss quadrature information for triangular regions:

$$\int_0^1 \int_0^1 f(L_1, L_2)\,dL_1\,dL_2 = \Sigma_{i=1}^n H_i f(L_{1i}, L_{2i})$$

Degree of polynomial	n	Coordinates L_1	L_2	Weighting function H_i
1	1	1/3	1/3	1/2
2	3	1/2	1/2	1/6
		1/2	0	1/6
		0	1/2	1/6
2	3	1/6	1/6	1/6
		2/3	1/6	1/6
		1/6	2/3	1/6
3	4	1/3	1/3	$-9/32$
		3/5	1/5	25/96
		1/5	3/5	25/96
		1/5	1/5	25/96
3	4	1/3	1/3	$-9/32$
		11/15	2/15	25/96
		2/15	2/15	25/96
		2/15	11/15	25/96
4	7	0	0	1/40
		1/2	0	1/15
		1	0	1/40
		1/2	1/2	1/15
		0	1	1/40
		0	1/2	1/15
		1/3	1/3	9/40

Evaluation of the element equations

The calculation of the element stiffness matrix proceeds in a similar way to the one-dimensional case previously considered. The basic equation is

$$[k] = t \int_A [B]^{\mathrm{T}}[D][B]\,dx\,dy$$

Using Eq. 8.44, this is written as

$$[k] = t \int_0^1 \int_0^1 [B(L_1 L_2)]^{\mathrm{T}}[D][B(L_1 L_2)]|J|\,dL_1\,dL_2 \qquad [8.45]$$

where the $[B]$ matrix is expressed in terms of the natural coordinates L_1

and L_2. However, because the inverse of the Jacobian might not be explicit, the following numerical integration is used:

$$[k] = t \sum_{i=1}^{n} H_i[B(L_{1i}L_{2i})]^{\mathrm{T}}[D][B(L_{1i}L_{2i})]|J(L_{1i}L_{2i})| \qquad [8.46]$$

where $B(L_{1i}L_{2i})$ is the value of the $[B]$ matrix at sampling point i. The number of sampling points n depends on the order of the element being used, as discussed previously. Therefore the evaluation of $[k]$ for the six-noded isoparametric triangular element requires the following steps:

(a) Selection of the number and position of the n integration points and weighting functions. Generally three sampling points are used with the quadratic triangular element, and seven sampling points with the cubic triangular element.
(b) For each of the n integration points with coordinates (L_{1i}, L_{2i}):
  (i)   calculation of $\partial N_\beta/\partial L_1$, $\partial N_\beta/\partial L_2$ for $\beta = 1$ to 6, from Eq. 8.43
  (ii)   calculation of $[J]$, $[J]^{-1}$ and $|J|$, as in Example 8.7
  (iii) calculation of $\partial N_\beta/\partial x$, $\partial N_\beta/\partial y$ for $\beta = 1$ to 6, from Eq. 8.40
  (iv)   calculation of $[B]$, for example from Eq. 8.38
  (v)   calculation of $tH_i[B(L_{1i}L_{2i})]^{\mathrm{T}}[D][B(L_{1i}L_{2i})]|J(L_{1i}L_{2i})|$ and summation to give $[k]$.

When these steps are compared with the trivial procedures involved in the calculation of the stiffness matrix for a simplex element (*i.e.* Eq. 6.82), the penalties of using such a complex element are obvious. The number of calculations required is increased significantly.

Example 8.8: using numerical integration in a two-dimensional element

Calculate the contribution that the sampling point at $L_1 = L_2 = 0.5$ makes to the element stiffness matrix for the thermal element shown in Fig. 8.13.

To calculate the stiffness matrix, the numerical integration outlined in Eq. 8.46 must be performed. Thus the $[B]$ matrix and the Jacobian $[J]$ must be evaluated at the requested sampling point.

   The element is the same as that used in Example 8.7, where the Jacobian and its determinant were calculated for the same point. The inverse of the Jacobian is then

$$[J]^{-1} = \frac{1}{8}\begin{bmatrix} -2 & 2 \\ -2 & -2 \end{bmatrix} = \begin{bmatrix} 0.25 & 0.25 \\ -0.25 & -0.25 \end{bmatrix} \qquad [8.47]$$

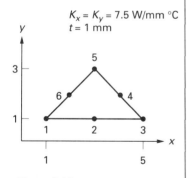

$K_x = K_y = 7.5$ W/mm °C
$t = 1$ mm

Figure 8.13

The [B] matrix for this element is defined by Eq. 8.38, where the derivatives of the shape functions with respect to the global coordinates are given by Eq. 8.40. Therefore, using the results of Example 8.7 and in particular Eq. 8.43, the derivatives of shape function N_1 at $L_1 = L_2 = 0.5$ are

$$\begin{Bmatrix} \dfrac{\partial N_1}{\partial x} \\ \dfrac{\partial N_1}{\partial y} \end{Bmatrix} = [J]^{-1} \begin{Bmatrix} \dfrac{\partial N_1}{\partial L_1} \\ \dfrac{\partial N_1}{\partial L_2} \end{Bmatrix} = [J]^{-1} \begin{Bmatrix} 4L_1 - 1 \\ 0 \end{Bmatrix}$$

$$= \begin{bmatrix} -0.25 & 0.25 \\ -0.25 & -0.25 \end{bmatrix} \begin{Bmatrix} 1 \\ 0 \end{Bmatrix} = \begin{Bmatrix} -0.25 \\ -0.25 \end{Bmatrix} \qquad [8.48]$$

If the other terms of the [B] matrix are calculated in the same way, then finally

$$[B(L_{1i}L_{2i})] = \begin{bmatrix} -0.25 & 0 & 0.25 & 0 & 0 & 0 \\ -0.25 & -1.0 & -0.25 & 1.0 & -0.5 & 1.0 \end{bmatrix}$$

$$[8.49]$$

The [D] matrix is

$$[D] = \begin{bmatrix} 7.5 & 0 \\ 0 & 7.5 \end{bmatrix} \qquad [8.50]$$

and the determinant of the Jacobian at the sampling point is 8, from Example 8.7. Hence from Eq. 8.46, using a weighting factor of 1/6, the sampling point specified is found to produce a stiffness matrix term of

$$\begin{bmatrix} 1.25 & 2.5 & 0.0 & -2.5 & 1.25 & -2.5 \\ 2.5 & 10.0 & 2.5 & -10.0 & 5.0 & -10.0 \\ 0.0 & 2.5 & 1.25 & -2.5 & 1.25 & -2.5 \\ -2.5 & -10.0 & -2.5 & 10.0 & -5.0 & 10.0 \\ 1.25 & 5.0 & 1.25 & -5.0 & 2.5 & -5.0 \\ -2.5 & -10.0 & -2.5 & 10.0 & -5.0 & 10.0 \end{bmatrix}$$

Similar terms can be derived for the other sampling points, and the complete stiffness matrix calculated by their summation.

The calculation of the rest of the element equations is not so difficult, and does not require numerical integration because the local coordinate integration formulae can still be applied. The format of the resulting terms, however, is unexpected. For example, consider the quadratic stress element with applied pressure shown in Fig. 8.14. The effect is included in the finite element equations with a term (as defined in Eq. 5.21) of the form

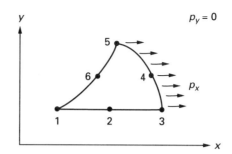

Figure 8.14 Stress element with distributed pressure load

$$\int_S [N]^\mathrm{T} \begin{Bmatrix} p_x \\ p_y \end{Bmatrix} \mathrm{d}S \qquad\qquad [8.51]$$

The shape function matrix for this element is given in Eq. 8.37. Since the pressure is applied down the sides with nodes 3, 4 and 5, then $L_1 = 0$ (as defined in Fig. 8.10), and consequently the three shape functions $N_1 = N_2 = N_6 = 0$ from Eq. 8.36. Therefore Eq. 8.51 becomes

$$\int_{S_{345}} \begin{bmatrix} 0 & 0 \\ 0 & 0 \\ 0 & 0 \\ 0 & 0 \\ N_3 & 0 \\ 0 & N_3 \\ N_4 & 0 \\ 0 & N_4 \\ N_5 & 0 \\ 0 & N_5 \\ 0 & 0 \\ 0 & 0 \end{bmatrix} \begin{Bmatrix} p_x \\ 0 \end{Bmatrix} \mathrm{d}S \qquad\qquad [8.52]$$

Here S is the area over which the pressure acts, but if the element has thickness t and the length of the side is H_{345} then the integral can be replaced by

$$\int_{H_{345}} \begin{Bmatrix} 0 \\ 0 \\ 0 \\ 0 \\ N_3 p_x \\ 0 \\ N_4 p_x \\ 0 \\ N_5 p_x \\ 0 \\ 0 \\ 0 \end{Bmatrix} t \, \mathrm{d}H \qquad\qquad [8.53]$$

The shape functions are defined in Eq. 8.36, and if they are substituted into Eq. 8.53 with the fact that $L_1 = 0$ then the local coordinate integration formulae can be used to readily evaluate the integral. Specifically, as in Eq. 4.25,

$$\int_H L_1^\alpha L_2^\beta \, dH = \frac{\alpha!\beta!}{(\alpha + \beta + 1)!} H$$

Therefore the first non-zero term will be

$$\int_{H_{345}} N_3 p_x t \, dH = p_x t \int_{H_{345}} L_2(2L_2 - 1) dH$$

$$= p_x t \int_{H_{345}} (2L_2^2 - L_2) dH$$

$$= p_x t \left(\frac{2}{3} - \frac{1}{2} \right) H_{345}$$

$$= \frac{1}{6} H_{345} p_x t$$

The integration of the other two terms yields

$$\int N_4 p_x t \, dH = \frac{2}{3} H_{345} p_x t$$

$$\int N_5 p_x t \, dH = \frac{1}{6} H_{345} p_x t$$

The final term is therefore

$$\frac{p_x H_{345} t}{6} \begin{Bmatrix} 0 \\ 0 \\ 0 \\ 0 \\ 1 \\ 0 \\ 4 \\ 0 \\ 1 \\ 0 \\ 0 \\ 0 \end{Bmatrix} \qquad [8.54]$$

Since $H_{345} p_x t$ equals the resultant load imposed on the face of the element, it is clear that one-sixth of the load is applied at the corner nodes while the remaining two-thirds is applied at the mid-side node, as shown in Fig. 8.15. The ratio of the forces is 1:4:1, which is by no means obvious. The same ratio is found when a quadratic thermal element is

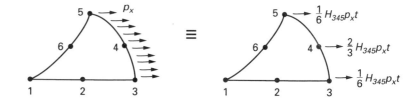

Figure 8.15 Nodal force representation of the applied pressure load

subjected to an incident heat flux. The mid-side node experiences four times the flux that the corner nodes experience.

If the body force vector is calculated in a similar way it leads to the following term:

$$\int_V [N]^T \left\{ \begin{array}{c} X \\ Y \end{array} \right\} dV = \frac{At}{3} \left\{ \begin{array}{c} 0 \\ 0 \\ X \\ Y \\ 0 \\ 0 \\ X \\ Y \\ 0 \\ 0 \\ X \\ Y \end{array} \right\}$$ [8.55]

Thus none of the element's mass is allocated to the corner nodes; it is equally divided between the mid-side nodes. Like the distribution of the pressure loading, this arrangement is not expected.

8.4.2 Isoparametric quadrilateral elements

The quadrilateral element is a multiplex element, and one of the most frequently used and reliable elements available (Fig. 8.16). The simplest element has four nodes, one at each corner, and has a linear interpolation function. The sides of the element must remain straight when in use. Quadratic and cubic versions of the element are also usually available with eight and twelve nodes respectively, and these can be used with curved sides. The natural coordinate system employed with this element is similar to the second coordinate system considered with the one-dimensional element in Section 8.3.3. For this two-dimensional element, the axes (ξ and η) are defined by lines drawn from the mid-side of opposite faces, as shown in Fig. 8.17, and the sides are defined by $\xi = \pm 1$ and $\eta = \pm 1$. Note that the element is distorted in Fig. 8.17 when in the global Cartesian system, but is perfectly square with sides parallel to the axes when referenced to the natural coordinate system.

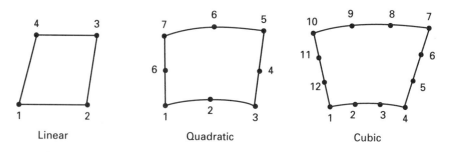

Figure 8.16 Quadrilateral family of elements

Linear   Quadratic   Cubic

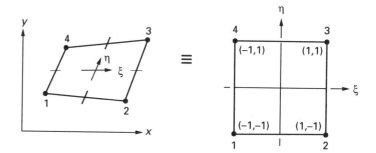

Figure 8.17 Natural coordinate system used in quadrilateral elements

Shape functions

The interpolation function required to define the linear element in Fig. 8.16 has four terms because the element has four nodes, and is as follows:

$$\phi = \alpha_1 + \alpha_2 \xi + \alpha_3 \eta + \alpha_4 \xi \eta \qquad [8.56]$$

This equation can also be expressed as

$$\phi = N_1 \phi_1 + N_2 \phi_2 + N_3 \phi_3 + N_4 \phi_4 \qquad [8.57]$$

where the shape functions are found to be

$$N_1 = \frac{1}{4}(1 - \xi)(1 - \eta) \qquad N_2 = \frac{1}{4}(1 + \xi)(1 - \eta)$$

$$[8.58]$$

$$N_3 = \frac{1}{4}(1 + \xi)(1 + \eta) \qquad N_4 = \frac{1}{4}(1 - \xi)(1 + \eta)$$

Figure 8.18 shows the variation of shape function N_1 over a typical element.

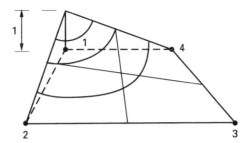

Figure 8.18 Variation of shape function N_1 over a quadrilateral element with linear interpolation function

The eight-noded quadrilateral element introduced in Fig. 8.16 will have interpolation function and shape functions defined as follows:

$$\phi = \sum_{i=1}^{8} N_i \phi_i \qquad [8.59]$$

where

$$N_1 = -\frac{1}{4}(1 - \xi)(1 - \eta)(1 + \xi + \eta) \qquad N_2 = \frac{1}{2}(1 - \xi^2)(1 - \eta)$$

$$N_3 = -\frac{1}{4}(1 + \xi)(1 - \eta)(1 - \xi + \eta) \qquad N_4 = \frac{1}{2}(1 + \xi)(1 - \eta^2)$$

$$N_5 = -\frac{1}{4}(1 + \xi)(1 + \eta)(1 - \xi - \eta) \qquad N_6 = \frac{1}{2}(1 - \xi^2)(1 + \eta)$$

$$N_7 = -\frac{1}{4}(1 - \xi)(1 + \eta)(1 + \xi - \eta) \qquad N_8 = \frac{1}{2}(1 - \xi)(1 - \eta^2)$$

$$[8.60]$$

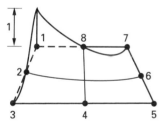

The variation of shape functions N_1 and N_2 over the element will follow the usual conventions as shown in Fig. 8.19. The geometry of these quadrilateral elements can also be expressed in a similar way to the field variable if the elements are isoparametric. For example, for the four-noded element,

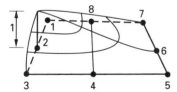

$$\left\{ \begin{array}{c} x \\ y \end{array} \right\} = \left[\begin{array}{ccccccc} N_1 & 0 & N_2 & 0 & \dots & N_8 & 0 \\ 0 & N_1 & 0 & N_2 & & 0 & N_8 \end{array} \right] \left\{ \begin{array}{c} x_1 \\ y_1 \\ x_2 \\ . \\ . \\ . \\ y_8 \end{array} \right\} \qquad [8.61]$$

Figure 8.19 Variation of shape functions N_1 and N_2 over a quadrilateral element with quadratic interpolation function

Coordinate transformation

Since the shape functions are defined as functions of the natural coordinates, a transformation must be devised to allow calculation of the [B] matrix, as was the case for the triangular element (Eq. 8.38). For example, consider shape function N_1:

$$\frac{\partial N_1}{\partial \xi} = \frac{\partial N_1}{\partial x}\frac{\partial x}{\partial \xi} + \frac{\partial N_1}{\partial y}\frac{\partial y}{\partial \xi} \qquad [8.62]$$

This is written as

$$\left\{\begin{array}{c} \dfrac{\partial N_1}{\partial x} \\[2mm] \dfrac{\partial N_1}{\partial y} \end{array}\right\} = [J]^{-1} \left\{\begin{array}{c} \dfrac{\partial N_1}{\partial \xi} \\[2mm] \dfrac{\partial N_1}{\partial \eta} \end{array}\right\} \qquad [8.63]$$

where the Jacobian is

$$[J] = \begin{bmatrix} \dfrac{\partial x}{\partial \xi} & \dfrac{\partial y}{\partial \xi} \\[2mm] \dfrac{\partial x}{\partial \eta} & \dfrac{\partial y}{\partial \eta} \end{bmatrix} \qquad [8.64]$$

Referring to Eqs 8.58, 8.60 and 8.61, it is clear that the Jacobian will be a function of ξ and η, and an explicit inverse might not exist.

Changing the variable of integration

Calculations of the element equations will require a change of integration variable, and the following relationship can be used:

$$\int_A f(x,y)\mathrm{d}x\,\mathrm{d}y = \int_{-1}^{1}\int_{-1}^{1} g(\xi,\eta)|J|\mathrm{d}\xi\,\mathrm{d}\eta \qquad [8.65]$$

where $g(\xi,\eta)$ is the function $f(x,y)$ written in terms of the natural coordinates ξ and η. As usual, this change of variable holds provided the determinant of the Jacobian [J] is greater than zero.

Numerical integration

Since the Jacobian is a function of the coordinates (ξ,η), integrations like that in Eq. 8.65 cannot be performed exactly, and numerical integration is required. The sampling points and weighting functions used for quadrilateral elements are shown in Table 8.3. A pattern of either 2×2 or 3×3 sampling points is used, depending on the order of the function to be evaluated. Generally the four-point formula is used for the four-noded quadrilateral, while the nine-point formula is used for the eight-noded version.

Table 8.3. Gauss quadrature information for quadrilateral regions:

$$\int_{-1}^{1}\int_{-1}^{1} f(\xi,\eta)\mathrm{d}\xi\mathrm{d}\eta = \Sigma_{i=1}^{n} H_i f(\xi_i,\eta_i)$$

	Degree of polynomial	n	Coordinates ξ	η	Weighting function H_i
	1	1	0	0	4.0
	3	4	$-1/\sqrt{3}$	$-1/\sqrt{3}$	1.0
			$1/\sqrt{3}$	$-1/\sqrt{3}$	1.0
			$-1/\sqrt{3}$	$1/\sqrt{3}$	1.0
			$1/\sqrt{3}$	$1/\sqrt{3}$	1.0
	4	9	$-\sqrt{3/5}$	$-\sqrt{3/5}$	25/81
			0	$-\sqrt{3/5}$	40/81
			$\sqrt{3/5}$	$-\sqrt{3/5}$	25/81
			$-\sqrt{3/5}$	0	40/81
			0	0	64/81
			$\sqrt{3/5}$	0	40/81
			$-\sqrt{3/5}$	$\sqrt{3/5}$	25/81
			0	$\sqrt{3/5}$	40/81
			$\sqrt{3/5}$	$\sqrt{3/5}$	25/81

Evaluation of the element equations

The calculation of the element equations proceeds in a similar way to the triangular element. However, unlike the triangular element, the option of local coordinate integration formulae is not available. Consequently the integrations must be evaluated precisely where possible, or alternatively numerical integration must be used.

The stiffness matrix is calculated using numerical integration according to Eq. 8.65:

$$[k] = t\int_A [B]^\mathrm{T}[D][B]\mathrm{d}x\,\mathrm{d}y$$

$$= t\int_{-1}^{1}\int_{-1}^{1} [B(\xi,\eta)]^\mathrm{T}[D][B(\xi,\eta)]|J|\mathrm{d}\xi\,\mathrm{d}\eta$$

Hence

$$[k] = t\sum_{i=1}^{n} H_i[B(\xi_i,\eta_i)]^\mathrm{T}[D][B(\xi_i,\eta_i)]|J(\xi_i,\eta_i)| \qquad [8.66]$$

The actual steps involved with this process are similar to those used in the evaluation of Eq. 8.46, except that the coordinates are ξ and η rather than L_1 and L_2.

The other terms of the element equations can be divided into those involving integration over a side, and those requiring integration over the face area of the element. For example, consider heat flux incident on side 3-4-5 of a quadrilateral element as shown in Fig. 8.20. The term that

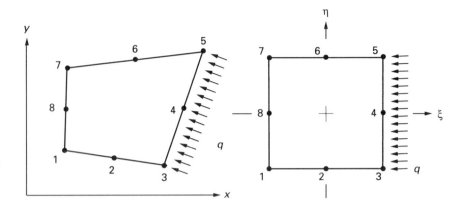

Figure 8.20 Heat flux incident on a quadrilateral element, in global and local coordinates

should be added to the force vector is

$$\int_S q[N]^{\mathrm{T}}\mathrm{d}S$$

When the element is referenced to the natural coordinate system, side 3-4-5 is seen to lie parallel to the η axis, at $\xi = 1$, where $N_1 = N_2 = N_6 = N_7 = N_8 = 0$. If the side has length H_{345} and thickness t, then the equation can be written as

$$\int_S q[N]^{\mathrm{T}}\mathrm{d}S = \frac{H_{345}t}{2}\int_{-1}^{1} q \left\{ \begin{array}{c} 0 \\ 0 \\ N_3 \\ N_4 \\ N_5 \\ 0 \\ 0 \\ 0 \end{array} \right\} \mathrm{d}\eta$$

$$= \frac{H_{345}t}{2}\int_{-1}^{1} q \left\{ \begin{array}{c} 0 \\ 0 \\ -\frac{1}{2}(1-\eta)(\eta) \\ (1-\eta^2) \\ -\frac{1}{2}(1+\eta)(-\eta) \\ 0 \\ 0 \\ 0 \end{array} \right\} \mathrm{d}\eta \qquad [8.67]$$

where the shape functions are obtained from Eq. 8.60 with $\xi = 1$, and the change of variable is carried out by Eq. 8.32. Integration of the three non-zero terms then yields

$$\frac{qH_{345}t}{6}\begin{Bmatrix} 0 \\ 0 \\ 1 \\ 4 \\ 1 \\ 0 \\ 0 \\ 0 \end{Bmatrix}$$ [8.68]

In other words, the flux is allocated in the ratio 1:4:1 to nodes 3, 4 and 5. This is the same distribution found for the quadratic triangular element (Eq. 8.54). If the eight-noded quadrilateral element is used for a stress analysis problem, it is found by the same approach that a pressure load will be applied to the nodes in the similar 1:4:1 ratio.

When body force terms are included in stress problems, the term

$$t\int_A [N]^T \begin{Bmatrix} \mathbb{X} \\ \mathbb{Y} \end{Bmatrix} dA$$ [8.69]

must be evaluated. The same term for higher-order triangular elements is easily evaluated by the natural coordinate integration formulae, but for quadrilateral elements the substitution of Eq. 8.65 must be applied to change the variable of integration. For a four-noded element it turns out that the body force is distributed equally amongst the nodes if the shape of the element is a parallelogram (Fig. 8.21(a)), but if the sides are not parallel then the distribution becomes more complex (Fig. 8.21(b)). It is interesting to note, however, that as the length of one side goes to zero, the distribution becomes that of the simplex triangle (Fig. 5.9).

When the body forces in eight-noded quadrilaterals are investigated by the same approach, the nodal allocation is unexpected (Fig. 8.22(a)): the corner nodes develop negative contributions. Again as the element is distorted from a parallel sided shape, the distribution varies. Also as the length of a side decreases, the forces approach those of a six-noded triangular element, which has the body force distributed equally between the mid-side nodes as shown in Fig. 8.22(b).

8.4.3 Isoparametric solid elements

A three-dimensional simplex element is introduced in Chapter 4, which has a tetrahedral shape. In practice, however, this element is rarely used

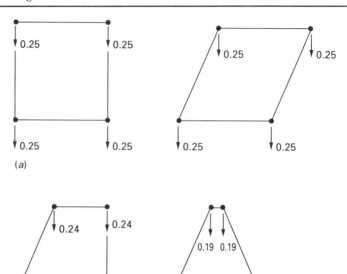

Figure 8.21 Distribution of a body force in the *y* direction for (a) parallel sided (b) general four-noded quadrilateral elements

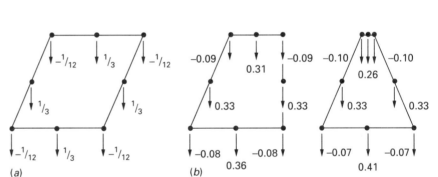

Figure 8.22 Distribution of a body force in the *y* direction for (a) parallel sided (b) general eight-noded quadrilateral elements

because it is difficult to generate models with such a shape, and it assumes a constant stress distribution. The brick elements shown in Fig. 8.23 are much more useful. It is easier to generate models with these shapes, and a stress gradient is allowed through each element. The linear element has 8 nodes, while the quadratic and cubic versions have 20 and 32 nodes respectively.

As with all the higher-order elements, the elements are defined in a natural coordinate system which allows them to be distorted in the global coordinate system, while remaining cube shaped in their own natural coordinates.

The implementation of the elements proceeds in an almost identical

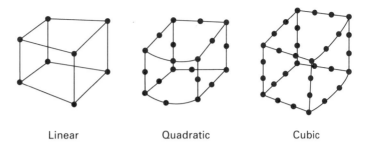

Figure 8.23 Hexahedral family
of elements

Linear   Quadratic   Cubic

way to the quadrilateral elements. For this reason, and because the
number of terms in the element matrices and vectors becomes so large,
they are not discussed in detail in this book. The basic linear element
when applied to stress problems results in a stiffness matrix of size
[24 × 24]. The Jacobian needed to evaluate the stiffness matrix is of size
[3 × 3], while the number of integration points for the linear, quadratic
and cubic elements is 8, 27 and 64 respectively (2^3, 3^3 and 4^3).

8.4.4   Stress and heat flow calculations

Solution of the finite element equations leads for example to a prediction
of the displacement or temperature distribution. Subsequent post-
processing of the results is then required to calculate the strains and
stresses or heat flows in each element. For elasticity problems, the strains
in an element are calculated from the displacements by

$$\{\varepsilon\} = [B]\{U\} \tag{8.70}$$

In simplex elements the matrix $[B]$ is a constant, and the strains are
constant. For the higher-order elements, however, $[B]$ is a function of the
coordinate system and not always explicit. Therefore the strains must be
evaluated at specific positions in such elements. The most obvious (and
accurate) locations are the sampling points used to calculate the $[B]$ and
$[k]$ matrices in the first place. The stresses of the nodes (particularly those
at the corners of the element) are more useful; however, to calculate a
nodal stress, either the $[B]$ matrix must be recalculated at the node for
use in Eq. 8.70, or the stress must be interpolated from the values at the
sampling points. In practice the second method is usually adopted, which
again adds further complexity to the implementation of these higher-
order elements.

8.5   Structural beam, plate and shell elements

The structural elements discussed in detail so far have translational
degrees of freedom, and therefore can only transmit forces directly from

element to element, with the nodes effectively acting as pin joints. There is, however, another very useful and important range of elements which can transmit not only forces but also moments, by specifying both translational and rotational degrees of freedom at their nodes. These elements are beams, plates and shells (curved plates).

Beams, plates and shells are often categorized as 'thick' or 'thin' depending on whether their formulations include transverse shear strains. For example, thick shells do allow consideration of the deformations due to shear and might be used in pressure vessels where the ratio of radius to wall thickness is less than ten. As with many aspects of the finite element method, a basic understanding of the engineering of the problem is essential for the correct selection and application of the element.

Beam element

Figure 8.24 shows a section of a (thin) beam undergoing simple bending. There is no deformation along the beam axis, and planes normal to the axis before loading remain normal after deformation (*i.e.* there is no shear deformation). The slope of the beam at any section is given by dv/dx, and the displacement in the x direction is found from

$$u = -y\frac{dv}{dx}$$
[8.71]

Therefore the bending strain in the beam is

$$\varepsilon_x = \frac{du}{dx} = -y\frac{d^2v}{dx^2}$$
[8.72]

The standard element which is used to model such a beam is shown in Fig. 8.25. It is developed with two nodes, and requires a cubic interpolation function for two reasons. Firstly, the element has four

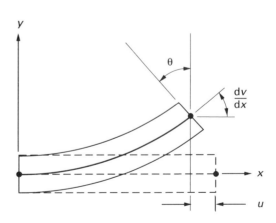

Figure 8.24 Deflection of a beam

boundary conditions (the four degrees of freedom), which demand four coefficients in the approximating function. In addition, however, to ensure that the element satisfies the necessary continuity conditions, that is deflection and slope continuity at the nodes, the interpolation function must be third order (see Section 8.6). The function is given by

Figure 8.25 Basic beam element

$$v = \alpha_1 + \alpha_2 x + \alpha_3 x^2 + \alpha_4 x^3 \qquad [8.73]$$

Note that the element is defined in a local coordinate system, and that the positive directions of the translations and rotations are marked on Fig. 8.25. Positive rotations are anticlockwise.

The α constants in Eq. 8.73 are evaluated in the usual manner by substitution of the boundary conditions for the element. These are

$$v = v_i \quad \text{and} \quad \theta = \frac{dv}{dx} = \theta_i \text{ at } x = 0$$

$$v = v_j \quad \text{and} \quad \theta = \frac{dv}{dx} = \theta_j \text{ at } x = L \qquad [8.74]$$

The resulting four equations can easily be solved to yield

$$\alpha_1 = v_i$$
$$\alpha_2 = \theta_i$$
$$\alpha_3 = \frac{3}{L^2}(v_j - v_i) - \frac{1}{L}(2\theta_i + \theta_j) \qquad [8.75]$$
$$\alpha_4 = \frac{2}{L^3}(v_i - v_j) + \frac{1}{L^2}(\theta_i + \theta_j)$$

When these are substituted back into the original interpolation function, it can be rearranged and expressed as

$$v = [N] \begin{Bmatrix} v_i \\ \theta_i \\ v_j \\ \theta_j \end{Bmatrix} = [N]\{U\} \qquad [8.76]$$

where

$$[N] = \left[1 - \left(\frac{3x^2}{L^2} + \frac{2x^3}{L^3} \right) x - \left(\frac{2x^2}{L} + \frac{2x^3}{L^2} \right) \left(\frac{3x^2}{L^2} - \frac{2x^3}{L^3} \right) - \left(\frac{x^2}{L} + \frac{x^3}{L^2} \right) \right]$$

[8.77]

As expected, the shape functions are third-order equations, the same as the interpolation function.

The stiffness matrix for the element is calculated from the potential energy in the usual way as described in Chapter 5. The strain energy is given by the standard equation

$$\Lambda = \int_V \frac{1}{2} \sigma_x \varepsilon_x \, dV = \frac{E}{2} \int \varepsilon_x^2 \, dV$$

From Eq. 8.72 this can be expressed as

$$\Lambda = \frac{E}{2} \int y^2 \left(\frac{d^2y}{dx^2} \right)^2 dV = \frac{E}{2} \int \left(\frac{d^2y}{dx^2} \right)^2 \left(\int y^2 \, dA \right) dx$$

But the area integral equals the second moment of area I; hence the strain energy is

$$\Lambda = \frac{EI}{2} \int_L \left(\frac{d^2v}{dx^2} \right)^2 dx$$

[8.78]

From Eq. 8.76, for one particular element,

$$\frac{d^2v}{dx^2} = \frac{d^2[N^{(e)}]}{dx^2} \{U^{(e)}\} = [B^{(e)}]\{U^{(e)}\}$$

[8.79]

where $[B^{(e)}]$ is simply calculated by the differentiation of the shape function matrix in Eq. 8.77. Thus

$$[B^{(e)}] = \left[\left(-\frac{6}{L^2} + \frac{12x}{L^3} \right) - \left(\frac{4}{L} + \frac{6x}{L^2} \right) \left(\frac{6}{L^2} - \frac{12x}{L^3} \right) - \left(\frac{2}{L} + \frac{6x}{L^2} \right) \right]$$

[8.80]

Substitution of Eq. 8.79 into Eq. 8.78 then gives

$$\Lambda^{(e)} = \frac{1}{2} \int_L \{U^{(e)}\}^T [B^{(e)}]^T [D^{(e)}][B^{(e)}]\{U^{(e)}\} dx$$

where $[D^{(e)}] = EI$.

Minimization of the potential energy then gives the now familiar form of the stiffness matrix,

$$[k^{(e)}] = \int_L [B^{(e)}]^T [D^{(e)}][B^{(e)}] dx$$

[8.81]

When the $[B]$ matrix is substituted into Eq. 8.81 and the integration carried out, the element stiffness matrix is found to be

$$[k^{(e)}] = \frac{EI}{L^3} \begin{bmatrix} 12 & 6L & -12 & 6L \\ 6L & 4L^2 & -6L & 2L^2 \\ -12 & -6L & 12 & -6L \\ 6L & 2L^2 & -6L & 4L^2 \end{bmatrix} \qquad [8.82]$$

The force vector for the element comprises solely the forces and moments applied at the nodes, namely

$$\{f^{(e)}\} = [S_i \ M_i \ S_j \ M_j] \qquad [8.83]$$

Thus

$$\frac{EI}{L^3} \begin{bmatrix} 12 & 6L & -12 & 6L \\ 6L & 4L^2 & -6L & 2L^2 \\ -12 & -6L & 12 & -6L \\ 6L & 2L^2 & -6L & 4L^2 \end{bmatrix} \begin{Bmatrix} v_i \\ \theta_i \\ v_j \\ \theta_j \end{Bmatrix} = \begin{Bmatrix} S_i \\ M_i \\ S_j \\ M_j \end{Bmatrix} \qquad [8.84]$$

The forces are labelled S_i and S_j because they are of course equal to the shear forces at the nodes, and indeed Eq. 8.84 can be used, after the displacements and rotations have been calculated, to determine the internal forces in the beam.

The following example illustrates the use of beam elements.

Example 8.9: beam analysis

Calculate the deflection under the load in the statically indeterminate beam in Fig. 8.26, and predict the shear force and bending moment distributions.

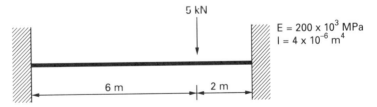

Figure 8.26

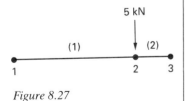

Figure 8.27

The finite element model for the beam need only consist of two elements, as in Fig. 8.27. The stiffness matrix for element (1) is

$$\frac{0.8 \times 10^{-6}}{6^3} \begin{bmatrix} 12 & 36 & -12 & 36 \\ 36 & 144 & -36 & 72 \\ -12 & -36 & 12 & -36 \\ 36 & 72 & -36 & 144 \end{bmatrix} \qquad [8.85]$$

while the stiffness matrix for element (2) is

$$\frac{0.8 \times 10^{-6}}{2^3} \begin{bmatrix} 12 & 12 & -12 & 12 \\ 12 & 16 & -12 & 8 \\ -12 & -12 & 12 & -12 \\ 12 & 8 & -12 & 8 \end{bmatrix} \qquad [8.86]$$

Including the load of 5 kN applied at node 2, the final set of system equations is then

$$10^{-6} \times \begin{bmatrix} 0.044 & 0.133 & -0.044 & 0.133 & 0.0 & 0.0 \\ 0.133 & 0.533 & -0.133 & 0.266 & 0.0 & 0.0 \\ -0.044 & -0.133 & 1.244 & 1.067 & -1.2 & 1.2 \\ 0.133 & 0.266 & 1.067 & 2.133 & -1.2 & 0.8 \\ 0.0 & 0.0 & -1.2 & -1.2 & 1.2 & -1.2 \\ 0.0 & 0.0 & 1.2 & 0.8 & -1.2 & 0.8 \end{bmatrix} \begin{Bmatrix} v_1 \\ \theta_1 \\ v_2 \\ \theta_2 \\ v_3 \\ \theta_3 \end{Bmatrix}$$

$$= \begin{Bmatrix} 0 \\ 0 \\ 5 \times 10^3 \\ 0 \\ 0 \\ 0 \end{Bmatrix}$$

However, before the equations can be solved, the constraint conditions must be applied, namely

$$v_1 = \theta_1 = v_3 = \theta_3 = 0$$

These result in a reduced set of system equations of

$$10^{-6} \times \begin{bmatrix} 1.244 & 1.067 \\ 1.067 & 2.133 \end{bmatrix} \begin{Bmatrix} v_2 \\ \theta_2 \end{Bmatrix} = \begin{Bmatrix} 5 \times 10^3 \\ 0 \end{Bmatrix}$$

Solution of these equations gives

$$v_2 = 0.007\ 037 \text{ m} \quad \text{and} \quad \theta_2 = -0.003\ 52 \text{ rad}$$

which compares with a theoretical value of 0.007 031 m for the vertical displacement.

When these results are substituted back into the element equations, as defined by Eq. 8.84, the shear force and bending moments in each element are found to be as follows:

Element	Node	Shear force (N)	Bending moment (N m)
1	1	−782.1	237.8
	2	782.1	−2815.6
2	2	4220.4	2812.4
	3	−4220.4	5628.4

All the results from this analysis agree closely with standard theoretical calculations, even though just two elements are used. This is because the cubic interpolation function predicts the actual deflection of the beam so accurately.

The beam element is defined in its own local coordinate system and is therefore of little use in its present form for real engineering problems. A coordinate transformation is required as described in Section 7.2 to allow the beam to be used in two- and three-dimensional space. In fact the same beam is examined in Example 7.1 and transformed into a two-dimensional form. The transformation matrix is found to be (Eq. 7.8)

$$[\lambda] = \begin{bmatrix} -m & l & 0 & 0 & 0 & 0 \\ 0 & 0 & 1 & 0 & 0 & 0 \\ 0 & 0 & 0 & -m & l & 0 \\ 0 & 0 & 0 & 0 & 0 & 1 \end{bmatrix}$$

where l and m are the direction cosines of the beam with respect to the global coordinate system (see Fig. 7.1), and the global stiffness matrix $[k^o]$ is calculated from

$$[k^o] = [\lambda]^T[k][\lambda]$$

The resulting matrix is of size $[6 \times 6]$ because there are now one rotational and two translational degrees of freedom at each of the two nodes. By a similar approach the beam can be applied in three dimensions, and further formulations of the beam element are also possible which include axial force and torsion capabilities.

Thin plate bending element

The simplest thin plate element is a triangle with three degrees of freedom at each node, one translation and two rotations, as shown in Fig. 8.28. As in the derivation of the beam element, normals to the plate's mid-surface are assumed to remain normal after deformation. Thus

$$\theta_x = \frac{\partial w}{\partial y} \quad \text{and} \quad \theta_y = \frac{\partial w}{\partial x} \tag{8.87}$$

with

$$u = -z\frac{\partial w}{\partial x} \quad \text{and} \quad v = -z\frac{\partial w}{\partial y} \tag{8.88}$$

If a cubic interpolation function is selected for the displacement, as in the

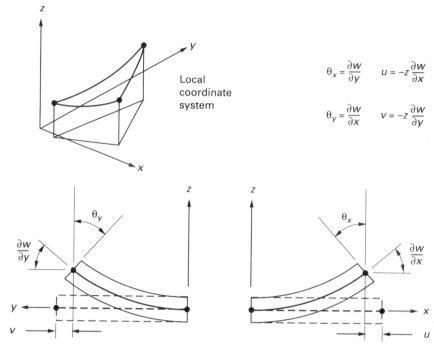

$$\theta_x = \frac{\partial w}{\partial y} \qquad u = -z\,\frac{\partial w}{\partial x}$$

$$\theta_y = \frac{\partial w}{\partial x} \qquad v = -z\,\frac{\partial w}{\partial y}$$

Figure 8.28 Thin plate bending element

beam element, then the displacement can be expressed as

$$w = \alpha_1 + \alpha_2 x + \alpha_3 y + \alpha_4 x^2 + \alpha_5 xy + \alpha_6 y^2$$
$$+ \alpha_7 x^3 + \alpha_8 (x^2 y + xy^2) + \alpha_9 y^3 \qquad [8.89]$$

(in the plate's local coordinate system). Nine coefficients are required because there are nine degrees of freedom, but note that the full cubic equation has ten terms, and consequently the α_8 term is mixed.

Unfortunately the plate element described is not ideal because, although the deflection and slopes of the plate are continuous at the nodes of the element, the slopes are not continuous along the element's sides. With such limitations the plate is said to be non-conforming, as described in Section 8.6. An equilateral triangle converges to the correct solution, but this is not true for less regularly shaped elements. A quadrilateral plate element can also be devised in a similar way, but again only the square form of the element converges to the correct answer. Despite the inaccuracies of these elements, they are capable of giving acceptable results and satisfactory accuracy in practical problems, and are therefore still used. Attempts have been and are still being made to improve the performance of this type of plate element.

Thick plate bending element

In the thick plate formulation the deflections and rotations are considered to be independent, and thus the normals (to the plate's mid-surface) do not remain normal, resulting in the development of transverse shear strains (Fig. 8.29). Therefore θ_x and θ_y are no longer functions of the out-of-plane deflection w and

$$u = -z\theta_y \quad \text{and} \quad v = -z\theta_x \qquad [8.90]$$

giving for example

$$\varepsilon_x = \frac{\partial u}{\partial x} = -z\frac{\partial \theta_y}{\partial x} \qquad [8.91]$$

and (a transverse shear strain)

$$\gamma_{zx} = \frac{\partial w}{\partial x} + \frac{\partial u}{\partial z} = \frac{\partial w}{\partial x} - \theta_y \qquad [8.92]$$

Consequently each node in the element has three independent degrees of freedom (in its local coordinate system), *i.e.* one translation and two rotations. Four- and eight-node versions of the plate are usually available,

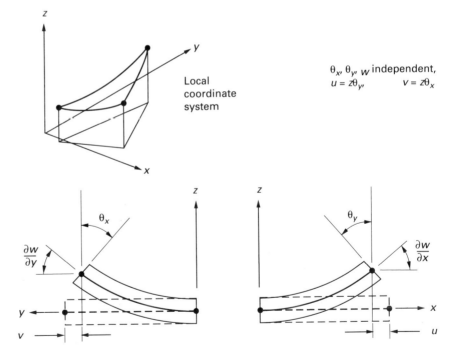

Figure 8.29 Thick plate bending element

the latter allowing the inclusion of curved sides. It turns out that this type of element is generally well behaved, and can even be used to model 'thin' plates with ratios of width to thickness of up to 50.

This thick plate element and the thin version described previously have no capability to support in-plane (membrane loads) in their current design. Since, however, the element is flat, and the bending and membrane effects are uncoupled (for small deflections), the latter effect can be included by simply superimposing a two-dimensional plane stress element formulation.

Shell elements

The difference between shell and plate elements is that the mid-surfaces of shells are curved, and as a result the membrane and bending effects are coupled, so that it is no longer possible to simply superimpose the two conditions. The formulation of shell elements is therefore more complex than that of flat plate elements, and is left to more specialized texts.

8.6   Convergence requirements of interpolation functions

So far convergence of the solution of finite element models has not been discussed, but it is a very important subject. As the size of the elements in a finite element analysis is reduced, so the solution of the model should converge to the exact answer, and interpolation functions should be chosen in order that this does occur. The easiest way to understand convergence requirements is to focus on a specific problem. For example, consider a two-dimensional stress analysis investigation. For convergence the following conditions must be met by the interpolation function:

(a)   The displacement must be continuous through each element.
(b)   As the size of an element decreases, the stresses in that element must approach constant values.
(c)   At the element boundaries, the displacement must be continuous. In other words, the elements must deform without causing openings or overlaps between adjacent elements.

A similar set of conditions can be devised for thermal elements, namely that the temperature distribution must be continuous in an element and must approach a constant value as the element size decreases, and adjacent elements must predict the same temperature for the same node.

Other problem types may require more demanding conditions. For example, beam elements require that not only the displacements but also the slopes of the displacements must be continuous across the element boundaries (see Section 8.5 and Example 8.9).

The conditions of convergence of a finite element analysis can be expressed in more general terms. For a particular finite element to converge to the correct answer as its size is decreased, its interpolation function must satisfy completeness and continuity conditions. These conditions relate to the order of the highest derivative of the field variable appearing in the functional for the particular problem. If the highest order is m, then:

(a)   For completeness, the field variable and its derivative up to order m should be able to assume a constant value in the element as the size of the element decreases.

(b)   For continuity, the field variable and its derivatives up to order $m - 1$ should be continuous at the element boundaries.

The completeness condition is necessary for convergence to the exact solution, but the continuity condition is not vital, providing that the errors introduced by the approximation decrease fast enough as the element size decreases. If an element satisfies both the completeness and continuity conditions it is said to be conforming or compatible, and convergence is guaranteed. If an element is complete and converges, but does not satisfy the continuity condition, it is called non-conforming or incompatible.

Example 8.10: element convergence criteria

The interpolation function for the beam element in Section 8.5 is

$$v = a_1 + a_2x + a_3x^2 + a_4x^3 \qquad [8.93]$$

which describes the vertical displacement at any position along the length of the element, as shown in Fig. 8.25. Each of the two nodes has two degrees of freedom, namely a vertical displacement and a rotation. Show that the element is conforming, remembering that the strain in the x direction in this beam element is given by

$$\varepsilon_x = -y\frac{\partial^2 v}{\partial x^2} \qquad [8.94]$$

The functional used in stress analysis problems is the potential energy, which equals the strain energy minus the work done by the loads. The strain energy for this beam element is

$$\Lambda = \frac{1}{2}\int_V \sigma_x\varepsilon_x \, dV = \frac{E}{2}\int_V \varepsilon_x^2 \, dV \qquad [8.95]$$

Thus the highest-order derivative will be that due to the strain, which is second order (Eq. 8.94). Hence, for the completeness

---

condition to be satisfied, v, $\partial v/\partial x$ and $\partial^2 v/\partial x^2$ must all be capable of assuming a constant value. Differentiation of Eq. 8.93 shows this is indeed possible.

For continuity at the element boundaries, v and $\partial v/\partial x$ must be continuous, which is the case, since v and θ $(= \partial v/\partial x)$ are the specified degrees of freedom at each node.

Since the element satisfies the completeness and continuity conditions, it is conforming.

8.7   Conclusions

The simplex elements introduced in Chapter 4 assume a linear variation of the displacement or temperature, whereas complex and multiplex elements assume quadratic, cubic and possibly higher-order variations. Clearly then more simplex elements are required to approximate a given distribution, particularly in areas with a steep change in the field variable. As has been seen in this chapter, the complexity of the calculations required in the derivation of the element equations is significantly increased with higher-order elements, and numerical integration is usually implemented. Despite the increased complexity of the calculations, higher-order elements are preferred, because of their increased accuracy and therefore reliability in modelling.

One of the disadvantages of the use of complex and multiplex elements compared with simplex elements is the significant increase in computer memory that they require. For example, the three-noded simplex and six-noded complex triangular elements develop stiffness matrices in stress problems of size $[6 \times 6]$ and $[12 \times 12]$ respectively. The latter is four times the size of the former. Thus although fewer higher-order elements may be needed to model a given problem, an overall increase in computer resources may still be required. However, because of the reduced number of elements, models using complex and multiplex elements can invariably be generated more quickly, saving personnel and computer time.

A further disadvantage of the higher-order elements is discovered when they are used to model a problem with a complicated geometry. A reduced number of complex elements might not be adequate to represent all of the local geometries of the object, whereas a large number of simplex elements could represent the geometry more accurately. This and other aspects of practical modelling are discussed in the following chapter.

Problems

8.1   Use the Gauss numerical integration method to evaluate the following integrals. Also, examine the effect of using a reduced integration scheme on the accuracy of the solution.

(a) $\displaystyle\int_{-1}^{1} (5x^5 + 3x^2 - 2)dx.$

(b) $\displaystyle\int_{0}^{1} (5x^5 + 3x^2 - 2)dx.$

(c) $\displaystyle\int_{0}^{1}\int_{0}^{1} (7x^4 + 2x^2y + 4y^3 - 3)dx\,dy$ (for a triangular region).

(d) $\displaystyle\int_{-1}^{1}\int_{-1}^{1} (6x^3 + xy^2 - y + 2)dx\,dy$ (for a quadrilateral region).

8.2   Prove that the interpolation function of the quadratic element shown in Fig. 8.5 is

$$\phi = \phi_1 + (-3\phi_1 + 4\phi_2 - \phi_3)|L_2 + 2(\phi_1 - 2\phi_2 + \phi_3)|L_2^2$$

8.3   Rearrange the equation in Problem 8.2 to derive the shape functions for the one-dimensional quadratic element.

8.4   Derive the shape functions for a one-dimensional element with a cubic interpolation function.

8.5   Calculate the $[B]$ matrix for a one-dimensional cubic element.

8.6   Evaluate the following integrals for a one-dimensional quadratic thermal element, considering convection from both the perimeter and the end of the element:

(a) $\displaystyle\int Q[N]^{\mathrm{T}}dV$

(b) $\displaystyle\int h\phi_\infty[N]^{\mathrm{T}}dS.$

8.7   Evaluate the following integrals for a one-dimensional cubic thermal element, where convection only occurs around the perimeter of the element:

(a) $\displaystyle\int h[N]^{\mathrm{T}}[N]dS$

(b) $\displaystyle\int h\phi_\infty[N]^{\mathrm{T}}dS$

(c) $\displaystyle\int Q[N]^{\mathrm{T}}dV.$

8.8   Show that the Jacobian for the one-dimensional quadratic element in Fig. 8.9 is $L/2$ when the mid-side node is half-way along the length of the element. Plot the variation of the Jacobian along the element as the position of the mid-side node varies.

8.9   Examine the convergence of the straight fin in Example 8.5 by comparing the results predicted by models using one, two, three and four elements.

8.10   Examine what happens to the determinant of the Jacobian of the element in Example 8.7 when node 2 is positioned at (2,0).

8.11   Evaluate the following integrals for a triangular quadratic thermal element, considering convection from both the edge and the face of the element:

(a) $\int Q[N]^{\mathrm{T}} \mathrm{d}V$

(b) $\int h\phi_\infty [N]^{\mathrm{T}} \mathrm{d}S.$

8.12   Evaluate the following integrals for a triangular cubic thermal element, considering convection from both the edge and the face of the element:

(a) $\int Q[N]^{\mathrm{T}} \mathrm{d}V$

(b) $\int h\phi_\infty [N]^{\mathrm{T}} \mathrm{d}S.$

8.13   Derive the force vector term required to account for the linearly varying pressure load shown in Fig. 8.30. [Hint: express the pressure distribution as a function of the nodal values using the shape functions.]

8.14   Calculate the force vector term for the elements in Fig. 8.31(a)–(d).

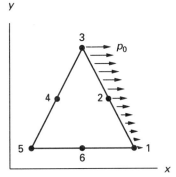

Figure 8.30

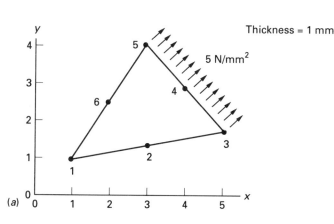

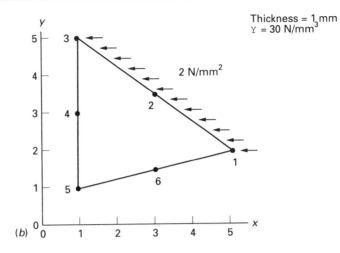

Thickness = 1 mm
Υ = 30 N/mm³

2 N/mm²

(b)

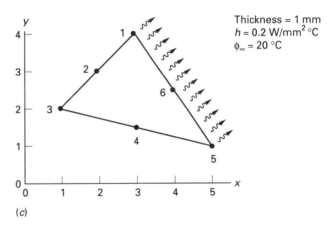

Thickness = 1 mm
h = 0.2 W/mm² °C
ϕ_∞ = 20 °C

(c)

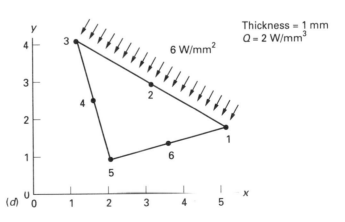

Thickness = 1 mm
Q = 2 W/mm³

6 W/mm²

(d)

Figure 8.31

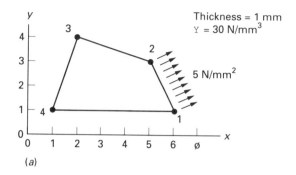

(a)

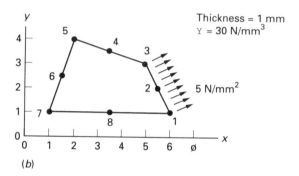

(b)

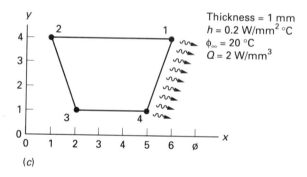

(c)

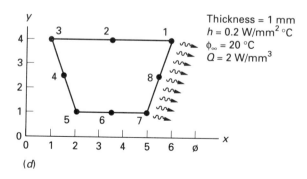

Figure 8.32   (d)

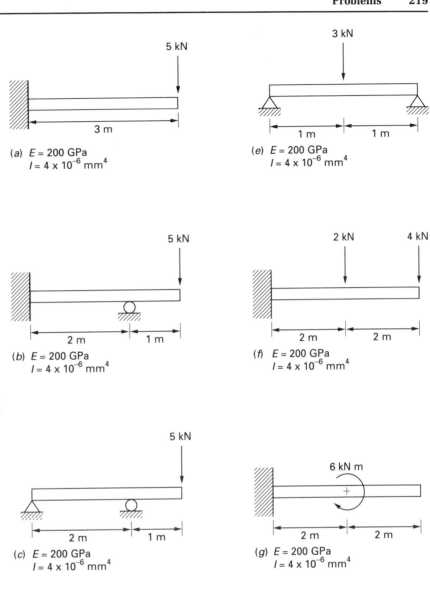

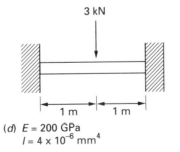

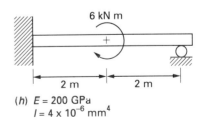

Figure 8.33

(a) E = 200 GPa
I = 4 x 10^{-6} mm^4

(b) E = 200 GPa
I = 4 x 10^{-6} mm^4

(c) E = 200 GPa
I = 4 x 10^{-6} mm^4

(d) E = 200 GPa
I = 4 x 10^{-6} mm^4

(e) E = 200 GPa
I = 4 x 10^{-6} mm^4

(f) E = 200 GPa
I = 4 x 10^{-6} mm^4

(g) E = 200 GPa
I = 4 x 10^{-6} mm^4

(h) E = 200 GPa
I = 4 x 10^{-6} mm^4

8.15 Derive the shape functions for a four-noded quadrilateral element.

8.16 Show that the field variable is continuous across the interelement boundary of two adjacent four-noded quadrilateral elements when natural (ξ, η) coordinate systems are used. [Hint: write down the interpolation functions for the two elements along the common boundary.]

8.17 Evaluate the following integrals for a four-noded quadrilateral thermal element, considering convection from both the edge and the face of the element:

(a) $\int Q[N]^T dV$

(b) $\int h\phi_\infty [N]^T dS.$

8.18 Calculate the force vector term for the elements in Fig. 8.32(a)–(d).

8.19 Derive the finite element equations for a thick beam. [Hint: apply the ideas of thick plate theory to the thin beam derivation.]

8.20 Analyse the beam problems in Fig. 8.33(a)–(h) using the standard thin beam formulation, and compare the results with those calculated theoretically.

9 Modelling procedures and results processing

9.1 Introduction

The finite element method is only approximate. The way in which the problem is modelled, for example the approximations in the geometry or material properties, and the method of discretization (*i.e.* the number and type of elements used), both affect the accuracy of the final answer. This chapter is concerned with modelling procedures and the use of finite element software to obtain accurate solutions. The accuracy of the implementation of the elements is discussed in Chapter 11.

After the importance of developing both a valid and an accurate representation of the problem is examined, element selection and mesh refinement are discussed in detail, and comparisons of the performance of different element types and mesh densities are reported. This is followed by a section which examines the effect of element distortion on the accuracy of the solution.

It will soon become obvious to the new user of finite element software that even the simplest of models produces a large amount of output data. For example, each node of each element in a three-dimensional stress analysis problem will provided a minimum of three displacements and six stress components. The processing and simplification of these results is an important step in the procedure and is also discussed in this chapter.

Finally, methods of checking the results of a finite element model are presented. This is vital in any analysis, and the results should not be relied upon without some form of verification.

9.2 Model validity and accuracy

A finite element analysis will only produce the correct answer if the model is both valid and accurate. Validity depends on how faithfully the physical problem is represented in the computer, while accuracy depends on how close the model is to convergence. Clearly the answer will only converge to the solution of the computer's representation of the problem.

For example, by using a plane strain model, the user will find the solution of a two-dimensional approximation, but not the original three-dimensional problem. If the elements are formulated correctly, then as the mesh density of a model increases, and thus the size of the individual elements decreases, the solution should converge to the exact answer. The necessary conditions for convergence, namely completeness and continuity of the interpolation functions, are discussed in Section 8.6. It should be obvious that it is not just the number of elements that is important when convergence is considered. The layout of the elements and the distortion of individual elements both significantly affect convergence rates, and these effects are discussed in later sections of this chapter.

Model validity

When modelling a complex problem, approximations might be made in one or more of the following:

(a)   geometry
(b)   material properties
(c)   loading conditions
(d)   constraint conditions.

The acceptability of those approximations depends on the purpose of the investigation, and indeed some comparative analyses do not require absolute accuracy. If simplifications in the model are necessary, then they should be examined somehow in an effort to quantify their effects. This is discussed again in Section 9.6.

Geometry

There are two levels of geometrical approximations that can occur in a model. At an element level, complex shapes or particularly small features in a body might not be represented precisely by the elements, since even the most advanced elements can only take up limited curved shapes. As already discussed in Section 8.1, complex and multiplex elements are superior in this respect because they can at least follow a polynomial shape. However, even these higher-order elements can only approximate the most common non-polynomial curve, a circular arc, as shown in Fig. 9.1. The maximum radial errors between the quadratic curve and the 90° and 45° arcs are 1 per cent and less than 0.1 per cent respectively, while the corresponding values for the simplex (straight line) versions are 29 per cent and 7.6 per cent. The advantage of using one complex element rather than two simplex elements is obvious. Note, however, that it is not

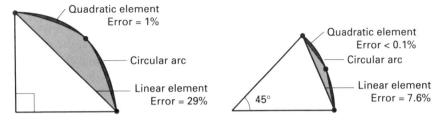

Figure 9.1 Linear and quadratic approximations to a circular arc, showing the maximum radial errors

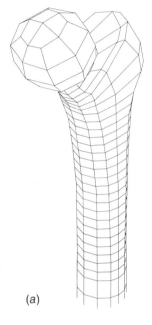

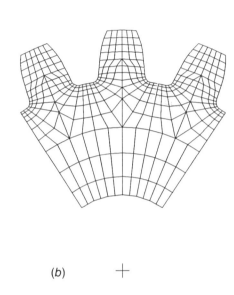

Figure 9.2 Models of (a) upper half of a human femur, with approximated geometry (b) part of a spur gear, to examine tooth stresses

(a)   (b)

using one complex element rather than two simplex elements is obvious. Note, however, that it is not advisable for a higher-order element to extend over more than 30° in areas where the stress gradient is expected to be high.

Geometrical approximations can be used legitimately to reduce the dimensionality of the problem, thus considerably simplifying the model as illustrated in Section 3.2, but it may also be necessary or desirable to approximate the overall geometry of the shape further. For example, the finite element model of the upper portion of a human femur in Fig. 9.2(a) does not represent the geometry of the bone exactly since the profile is very complex. With the other known approximations in the model, particularly the material properties and loading conditions, the time spent to develop a precise copy of the bone is not warranted. Figure 9.2(b) shows a model of a spur gear. If the purpose of this analysis is to examine the tooth stresses under normal loading conditions, it is not necessary to model all of the gear, since the rest of the structure will have little effect on the particular tooth stresses.

Material property

Material properties must frequently be approximated in finite element analyses. Indeed of course there is an accepted degree of variation in the Young's modulus of even the most standard of engineering materials. The properties of all materials vary with temperature, but fortunately the temperature range of most analyses is small enough to permit constant properties to be used. Many commercial packages, however, allow the user to input the variation of material properties with temperature. Not only does this mean that analyses can be readily performed at different temperatures, but more importantly any temperature variation through the problem can be reflected in the material properties. For example, thermal stress problems are quite common where a structural solution depends on the temperature distribution. For such a problem the temperature field is solved initially for the given thermal boundary conditions, and the resulting nodal temperatures are fed into the stress model to allow calculation of the relevant material properties.

Non-linearities in the material properties, for example plasticity and creep, are more complex. Such analyses depend on the actual load history applied to the model and require an incremental solution technique. An introduction to non-linear problems is included in Chapter 10.

Material property approximations are particularly common for orthotropic and anisotropic materials, where a material's performance depends on its orientation, as for example in laminated glass and carbon fibre reinforced plastics (Fig. 9.3). Inclusion of anisotropy in the finite element method is not difficult; the material property matrix $[D]$ (Eqs 5.75 and 6.10) must be adjusted accordingly. The usual difficulty, however, is that the complete set of material constants is not known, or the values are variable, for example with strain rate. Note that an anisotropic material

Figure 9.3 Layered composite material with anisotropic material properties

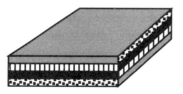
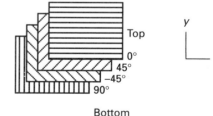

for an elasticity analysis requires 21 independent material properties, and an orthotropic material requires 9 as follows:

$$
\begin{Bmatrix} \varepsilon_x \\ \varepsilon_y \\ \varepsilon_z \\ \gamma_{xy} \\ \gamma_{yz} \\ \gamma_{zx} \end{Bmatrix} = \begin{bmatrix} \dfrac{1}{E_x} & \dfrac{-\nu_{xy}}{E_y} & \dfrac{-\nu_{xz}}{E_z} & 0 & 0 & 0 \\[2mm] \dfrac{-\nu_{yx}}{E_x} & \dfrac{1}{E_y} & \dfrac{-\nu_{yz}}{E_z} & 0 & 0 & 0 \\[2mm] \dfrac{-\nu_{zx}}{E_x} & \dfrac{-\nu_{zy}}{E_y} & \dfrac{1}{E_z} & 0 & 0 & 0 \\[2mm] 0 & 0 & 0 & \dfrac{1}{G_{xy}} & 0 & 0 \\[2mm] 0 & 0 & 0 & 0 & \dfrac{1}{G_{yz}} & 0 \\[2mm] 0 & 0 & 0 & 0 & 0 & \dfrac{1}{G_{zx}} \end{bmatrix} \begin{Bmatrix} \sigma_x \\ \sigma_y \\ \sigma_z \\ \tau_{xy} \\ \tau_{yz} \\ \tau_{zx} \end{Bmatrix}
$$

$$
= [D]^{-1}\{\sigma\} \qquad\qquad\qquad\qquad\qquad [9.1]
$$

A general three-dimensional anisotropic stress element will predict three direct and three shear stresses at a number of locations through the element (usually the integration or node points). For composite materials such as that shown in Fig. 9.3, this is generally of little use since the engineer requires the stresses in each layer. Because of the increasing importance of composite materials, many commercial finite element packages now include dedicated composite elements. With these elements, the user specifies the properties of the individual layers of the composite (typically up to 32), and then the program calculates the equivalent material property matrix [D]. After the strain in the element has been calculated, the detailed lay-up information then allows the program to calculate the stresses on the individual layers of the composite. Some software even predicts the failure criteria values (for example, maximum strain, maximum stress and Tsai-Wu) if the user supplies the necessary failure stresses.

Loading conditions

The finite element method converts all pressure and body forces to nodal values in stress problems, and all incident heat flux and heat sources to nodal values in thermal problems, as the derivations in Chapters 5, 6 and 8 clearly demonstrate. The application of pressures (for example) to simplex elements turns out to be trivial, but for complex elements the equivalent nodal distribution is by no means obvious (see Fig. 8.15). When a constant pressure is applied over a section of the model, it appears just as easy to specify the nodal forces directly rather than enter pressure values. However, this can quite easily lead to inaccuracies in the modelling. Consider the problem shown in Fig. 9.4. At first sight it might seem reasonable that the uniform pressure in Fig. 9.4(a) should be simply divided equally amongst the four nodes of the three elements as shown in Fig. 9.4(c), but this is incorrect. Instead, each node of each element should receive $pL/2$, so that the two internal nodes experience pL in total (Fig. 9.4(b)).

A similar model with quadratic elements is more complicated. Using the results of Chapter 8, and in particular Eq. 8.68, the distribution of a uniform pressure over three such elements is shown in Fig. 9.5.

Body force (and heat source terms in thermal analyses) are similarly combined in a cumulative way at common nodes.

In reality all loads (and heat fluxes) are applied over a finite region, although in many cases we assume a point contact. In the finite element method, applying a single point load will cause unreliable results in the immediate vicinity of the load, and ideally such a concentrated load should be input over the side of at least one element. (Naturally this does not apply to one-dimensional (line) elements which do handle single point loads exactly.) If the user is not interested in the stresses close to the point contact, and is only concerned with distant effects, then a single

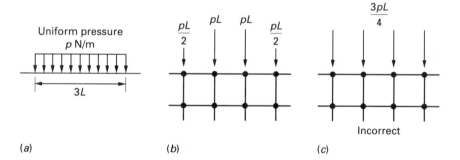

Figure 9.4 Application of a constant pressure over three linear elements

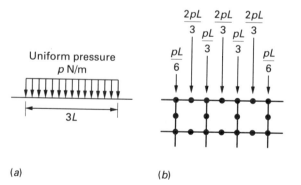

Figure 9.5 Application of a constant pressure over three quadratic elements

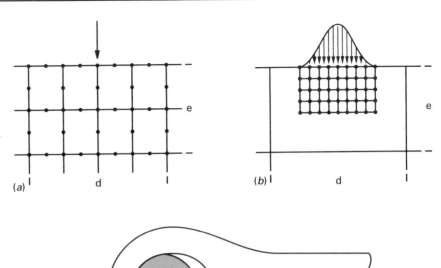

Figure 9.6 (a) Coarse mesh, for use where distant effects are required (b) fine mesh, for use where local effects are required

Figure 9.7 Pinned connection under axial loading

nodal load can be used (Fig. 9.6(a)). However, where local stresses are of primary importance, a fine mesh should be used, with the load distributed over the elements in a representative manner, as shown in Fig. 9.6(b). As a guide, the error in using a concentrated load can be assumed to be acceptable when the distance from the load is at least three and preferably five times the width of the original distributed load.

The forces applied to a model under investigation are caused by action and reaction with other bodies, and for certain problems the extent and magnitude of the contact effect can be difficult to determine. For stress analyses, these contact conditions can be awkward to model, and it is frequently necessary to include the contacting body to ensure the load is applied correctly. For example, to analyse the lug shown in Fig. 9.7 it is necessary to represent accurately the loading conditions, which are applied through the pin. The lug clearly experiences some form of pressure distribution on its internal surface, but the precise distribution will depend on the load, the clearance at the hole and the material properties of the two parts. To model the problem in the way shown in Fig. 9.8 with standard plane stress elements is incorrect because the model assumes a perfect connection between the pin and the lug – equivalent to the pin being glued in position.

Ideally a row of gap elements should be used to separate the pin and the lug. A gap element is a special element that most commercial

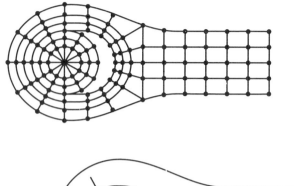

Figure 9.8 Plane stress representation of the problem with the pin connected to the lug

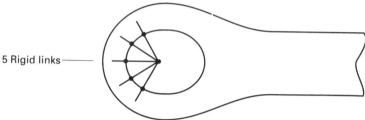

5 Rigid links

Figure 9.9 Modelling of the pin with rigid links

packages include, and is designed so that it transmits compressive but not tensile loads. A problem including gaps is solved iteratively, with the program finding the values of the gap displacements and the control forces which solve the problem. The cost of performing an analysis with gap elements can be significantly more than a simple static analysis, since the problem must be solved several times. (Note that, because of the way pressures are applied to the sides of quadratic and cubic elements, gap elements should not be applied to elements with mid-side nodes.)

If a highly accurate solution is not required, a much cheaper and usually acceptable alternative is to represent the pin with rigid links, as shown in Fig. 9.9. The links are simply pin-jointed elements that connect the nodes around the hole to a single node placed at the centre of the pin.

The axial stress distributions in the lug for the three modelling alternatives are presented in Plate 2. The analysis times for the different models are also compared, but note that this does not include the time to generate the models.

Constraint conditions

Since the constraints applied to a finite element model develop reaction forces, they can cause similar effects to those produced by the applied loading forces. In particular, a single point constraint may well lead to high localized stresses which result in misleading stress values. This is especially obvious when the results of such an analysis are presented as a contour plot by a post-processor. The ranges of the contours are automatically determined from the maximum and minimum values. If the maximum is localized and high, all the contours appear focused around the point, with the rest of the object appearing uniformly stressed. A more representative distribution can be obtained by specifying a maximum contour value, so that the localized stresses are ignored, or by plotting the stresses in a limited number of elements.

However, to overcome the problem completely the constraint can be distributed in the same way that a concentrated load is distributed. When an element face is being constrained, then all the nodes on the face, including any mid-side nodes, should be constrained. Note though that two or more inappropriate constraints can themselves lead to a high localized stress if they do not represent the nature of the boundary condition precisely. If the stresses are not required in the immediate vicinity of the constraint, then a point constraint will suffice.

The way in which constraints are applied to a model can alter the analysis from one problem to another. Consider for example pure tension of a bar and the finite element models of the problem in Fig. 9.10. Since the problem is symmetric, only half of the bar is modelled, with the symmetry being accounted for by the applied boundary conditions. For the specified loading, the constraint conditions shown in Fig. 9.10(a) should be applied. The x constraints prevent the plane of symmetry from moving in the x direction, while the single y constraint restrains the centre-line from moving vertically. A common mistake is to fix all the nodes or the plane in the y direction as illustrated in Fig. 9.10(b). This prevents the bar from contracting along that plane, and thus transverse

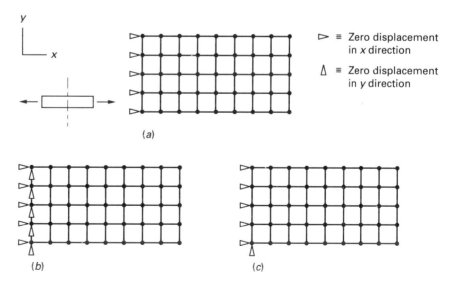

Figure 9.10 Same finite element model with different applied boundary conditions

stresses are induced. No transverse stresses are induced in pure tension, and in fact the model represents a cantilever bar rigidly fixed at its end. Figure 9.10(c) is also incorrect. The single y constraint forces the bar to move about that point rather than the centre-line. The deflections predicted by the three models are compared in Plate 3.

The layout of a finite element mesh can also produce different results with the same set of constraint conditions. For example, consider the two models of a square plate in Fig. 9.11. At a first glance the meshes appear equivalent. They both contain eight equally sized triangular elements, and all the edges are rigidly constrained. However, careful inspection of Fig. 9.11(a) shows that the corner elements are completely restrained, since all of their three nodes are fixed. Thus the model is really a representation of a smaller plate consisting of just the four inner elements. Figure 9.11(b) shows a preferred mesh layout.

Imposed displacements are rarely encountered, but when they are applied to a finite element model they must be specified with a very high degree of accuracy, because even the smallest relative displacements will cause large stresses.

9.3   Mesh design and refinement

The finite element modeller must have a sound understanding of the physics of a problem to allow him to design a representative and efficient finite element mesh. Analysing a problem by the use of a constant mesh density is attractive from the point of view of generating the mesh, but very inefficient and wasteful of resources. The mesh should reflect the expected changes in the field variable. For example, for an elasticity problem the user should estimate which areas will experience high stress gradients and which will be uniformly stressed, and design the mesh accordingly.

Figure 9.11 Different meshes of the same problem with the same boundary conditions

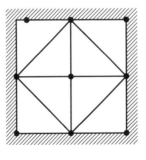

(a)

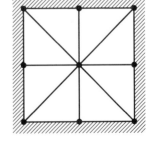

(b)

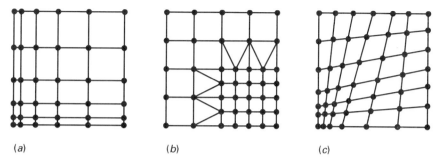

Figure 9.12 Methods of varying mesh density in two dimensions   (a)   (b)   (c)

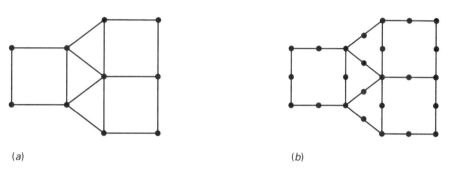

Figure 9.13 Acceptable mixing of elements: (a) simplex and multiplex elements (b) complex and multiplex elements   (a)   (b)

The sensitivity of a mesh can be varied in two ways. Either the mesh density can change from one part of the model to another, or the (interpolation) order of the elements can be varied. In a well constructed finite element model there may be several changes in mesh density and/or in the degree of complexity of the elements. Examples of methods of varying mesh density are introduced in Section 3.6; however, these are primarily one-dimensional in nature. Varying the mesh in two and three dimensions can be difficult, particularly if close attention is paid to minimizing element distortion. Figure 9.12 shows examples of methods of changing mesh density in two dimensions.

Note that one of the meshes uses both rectangular and triangular elements. This is quite acceptable providing that continuity is maintained at the element boundaries, that is, the field variable has the same form of distribution along the edges of connecting elements. Figure 9.13 shows mixing of simplex and first-order multiplex elements, and complex and second-order multiplex elements. It should be clear that elements with mid-side nodes should only be connected to other elements with mid-side nodes (unless special transition elements are available).

The element combinations shown in Fig. 9.14 are *not* advised. The mesh in Fig. 9.14(a) connects an edge with a linear variation to one with a quadratic variation. Since the side of the linear element will remain straight, a gap could open up in the model as illustrated. A similar effect

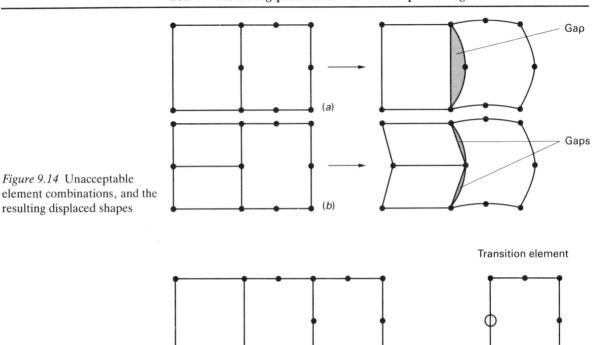

Figure 9.14 Unacceptable element combinations, and the resulting displaced shapes

Figure 9.15 Use of a transition element

will be seen with two linear elements connected to the multiplex element, as in Fig. 9.14(b). Although all the nodes are connected, the displacements still lead to gaps in the model.

If it is necessary to connect elements with different order interpolation functions, either transition elements must be used or a constraint equation must be imposed on the mid-side node(s). A transition element is shown in Fig. 9.15. It is used to join a first-order multiplex element to a second-order element. The implementation of such an element is not that difficult. In this example the transition element can simply be an eight-noded rectangle which forces the mid-side node to lie on a straight line between the corner nodes. This ensures that no gap opens up in the model and continuity is maintained. Such transition elements are usually available in commercial packages, and the user is unaware of the extra calculation involved. However, because the program imposes a certain displacement on the mid-side node, these elements should not be used in parts of the model which are of particular interest. A constraint equation has precisely the same effect as the 'unused' mid-side node, except that the user specifies the relationship between the mid-side and corner nodes manually.

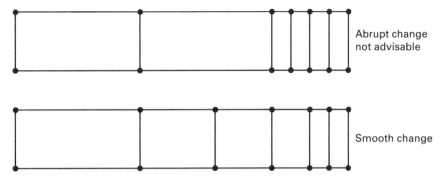

Figure 9.16 Variation from a coarse mesh density to a fine mesh density should be gradual

In areas of the model where a coarse mesh must be refined to a fine mesh or vice versa, an abrupt change in the element size should be avoided, since any errors in the coarse elements will be transferred into the fine mesh. The change from coarse to fine should be graduated, as illustrated in Fig. 9.16.

Significantly different sized elements should be avoided in a finite element model, as the resulting differences in the element stiffness matrices can lead to ill-conditioning of the global stiffness matrix. Since the element matrices are functions of the dimensions of the elements, very large and very small elements will lead to a large difference in the order of the terms in the stiffness matrices. Round-off errors, particularly on smaller (low-precision) computers, can then become significant and result in an unreliable solution. Most programs check for ill-conditioning by comparing the largest and smallest terms in the stiffness matrix, but other indications of less serious discrepancies can also be detected in the results as discussed in Section 9.6. As a rough guide, in the absence of more detailed guidelines, element lengths should not vary by more than 100:1 in a model, *i.e.* the ratio of areas should not exceed 10 000:1.

Large differences in material properties can also lead to problems in the stiffness matrix for similar reasons to the size effects. Again the ratio of terms in the stiffness matrix will become unsatisfactory, leading to possible ill-conditioning of the equations.

The way in which a finite element mesh is constructed can significantly alter the results produced from an analysis. The two meshes in Fig. 9.17 have exactly the same number, type and size of elements. They only differ in the way that the elements are laid out, and yet the mesh in Fig. 9.17(b) will produce more accurate results. This is because the first mesh does not repeat the axes of symmetry present in the problem, whereas the second mesh does.

So far in this section the discussion has concentrated on the basic 'solid' element types, such as those for analysing two-dimensional plane stress or

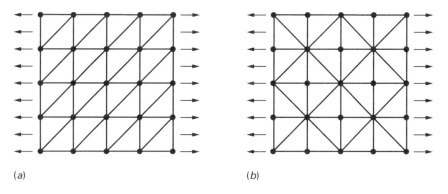

Figure 9.17 Two apparently similar meshes, but mesh (b) is more accurate

(a)   (b)

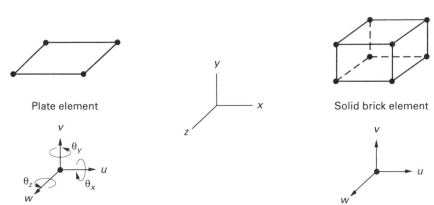

Figure 9.18 Nodal degrees of freedom used in plate and solid elements

strain problems. However, there are a number of other more sophisticated element types available to the user which can be used in conjunction with these standard elements, if care is taken with the development of the model. For example, most commercial packages include beam and plate elements, which are formulated to resist bending as well as tensile loads (see Section 8.5). The most general form of these beam and plate elements in three-dimensional space has six degrees of freedom per node, that is three translations and three rotations, while a three-dimensional solid element has three (translational) degrees of freedom per node (Fig. 9.18). Thus the nodes of the solid element types are essentially pin-jointed, and cannot transmit any rotational effects. Because of the rotational degrees of freedom of the plate elements, they can be used to model out-of-plane effects.

It is quite conceivable that a problem requires the use of both solid elements and plate elements, and indeed they can be used together if the different nodal degrees of freedom are taken into account. For example, in Fig. 9.19(a) the plate (p) is connected to the solid region (s). If the plate elements are simply connected to the edge nodes of the solid part,

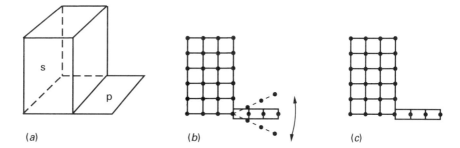

Figure 9.19 Connection of
plate and solid elements

(a)   (b)   (c)

then the connection will be pin-jointed, allowing rotation about the edge
(Fig. 9.19(b)). This can be overcome in two ways. Constraint equations
can be supplied by the user to relate the rotational degrees of freedom to
the translations of the adjoining solid elements. Alternatively, an
approximate but usually satisfactory answer is to continue the plate
elements over the bottom face of the solid region, as shown in Fig.
9.19(c). It is advisable to extend the plate elements over at least two solid
elements as shown in the figure, but by adopting this technique the plate
and brick elements can easily be connected. It is possible to mix other
elements in the same way, providing that any nodal incompatibilities are
treated with equal care.

Mesh refinement

The accuracy of a model can be improved in two ways. Either the
geometry can be divided into smaller elements, so that the mesh density
is increased, or the accuracy of the elements themselves can be improved
by using higher-order interpolation functions. These two techniques to
improve a model's accuracy are known as h-refinement and p-refinement,
respectively. The latter appears the simpler of the two methods for the
user of a finite element package, but in practice commercial programs
only offer linear and quadratic (and occasionally cubic) elements, so that
the opportunity for p-refinement is limited. Also, increasing the element
order from linear to quadratic, for example, leads to a significant increase
in the computer time needed to analyse the structure. To confirm the
convergence of a model by progressive h-refinement of the mesh, all
previous meshes should be contained in the finer meshes. This is
generally referred to as a reducible net, and an example of such a mesh is
shown in Fig. 9.20.

As an example of increasing mesh density, consider the analysis of the
cantilevered beam presented in Fig. 9.21. This problem is frequently
analysed, but is a very convenient example to illustrate and compare
the performance of finite elements. The beam can be assumed to be

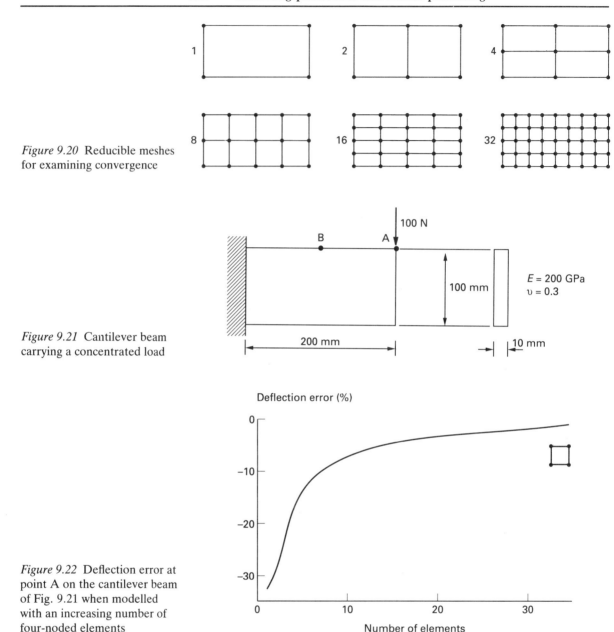

Figure 9.20 Reducible meshes for examining convergence

Figure 9.21 Cantilever beam carrying a concentrated load

Figure 9.22 Deflection error at point A on the cantilever beam of Fig. 9.21 when modelled with an increasing number of four-noded elements

two-dimensional, and a classical solution is readily available. The error in the deflection of point A is examined in Fig. 9.22 when the beam is modelled with linear four-noded rectangular elements. The error decreases as the number of nodes (and elements) increases, as expected. (Note, however, that just because the deflection at one point on the end of the beam has converged, it does not mean that the whole model has converged.) The stresses do not converge as quickly as the displacements, because they are not approximated to the same degree in the elements. The convergence of the bending stress at point B (half-way along the beam) is shown in Fig. 9.23.

If the cantilevered beam is examined with elements using different interpolation functions, then the advantages of the higher-order elements are immediately obvious. For example, Fig. 9.24 shows the results with linear, quadratic and cubic interpolation functions using two-, four- and eight-element meshes.

It must be remembered that the higher-order elements do require

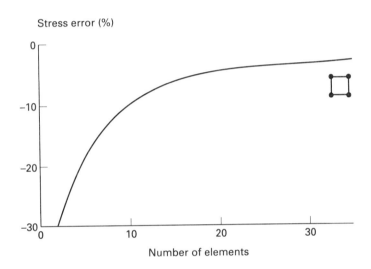

Figure 9.23 Bending stress at point B on the cantilever beam of Fig. 9.21 when modelled with an increasing number of four-noded elements

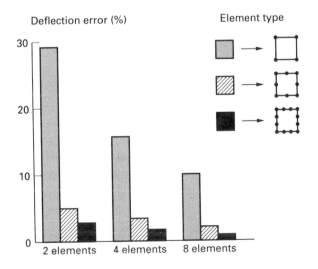

Figure 9.24 Deflection error at point A on the cantilever beam of Fig. 9.21 when modelled with elements of increasing complexity

significantly more computer time and resources for their solution. As an indication of these extra costs, the solution times for the series of models presented in the previous figures are summarized in Fig. 9.25. The times have been normalized for comparison purposes by assuming that the time for the simplest linear mesh is one unit. The times to solve the quadratic and cubic meshes are significantly greater than for the linear models. However, if models with equivalent errors are compared, the costs are not quite so different. For example, the two-element cubic mesh gives similar displacement results to the sixteen-element linear mesh, and the ratio of the costs is 1.5:1.6 (cubic:linear).

The simple beam bending problem is also ideal for comparing the

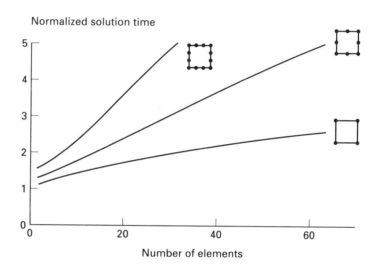

Figure 9.25 Solution times for various types and numbers of quadrilateral elements

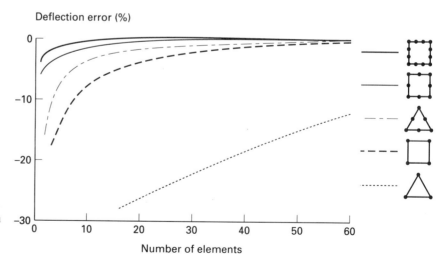

Figure 9.26 Deflection error at point A on the cantilever beam of Fig. 9.21 for different types and numbers of elements

performance of different element types. Both triangular and quadrilateral elements are represented in Fig. 9.26. It is not possible to draw definite guidelines from the results of just one problem, but it is clear that the six-noded triangle and eight-noded quadrilateral are significantly better than the simplex triangle and four-node quadrilateral.

9.4   Element distortion

The finite element method approximates the distribution of a variable through a component by assuming a known distribution of that variable in each element. However, as the elements become distorted, so errors are introduced into the elemental approximations. This section is concerned with the type and effects of element distortions, and discusses the checks that most commercial programs make on the user's model to identify any badly shaped elements. Guidelines on good element shapes are also introduced, but it is difficult to produce definitive rules since the effects of element distortion depend on many factors. Higher-order elements are less sensitive to element distortion, and the degree of allowable distortion depends on the stress (or heat flow) gradient experienced by the element. The smaller the gradient, the more the element can be distorted.

Element distortions relate to the skew, taper, warpage and aspect ratio of the faces and sides of the elements. In addition, incorrect positioning of the mid-side nodes of the higher-order elements and unreasonable internal angles of the elements indicate distortion. Examples illustrating the common forms of element distortion are shown in Fig. 9.27; where appropriate they can also be referred to element types other than those shown. Clearly an element may well exhibit more than one of these indications when it is distorted.

The type and number of element checks offered by finite element packages vary considerably from one package to another. The most common checks are based on

(a)   aspect ratio
(b)   internal angles
(c)   warpage angle
(d)   mid-side node position
(e)   value of the Jacobian.

The aspect ratio of an element is the ratio of the length of the largest side of the element to the shortest. For rectangular elements it is easy to observe the aspect ratio, but for quadrilateral elements which are skewed or tapered, and for triangular elements, calculation of the aspect ratio is not so obvious. The examples shown in Fig. 9.28 show the ways in which the ratio is usually calculated. It turns out that an equilateral triangle (*i.e.*

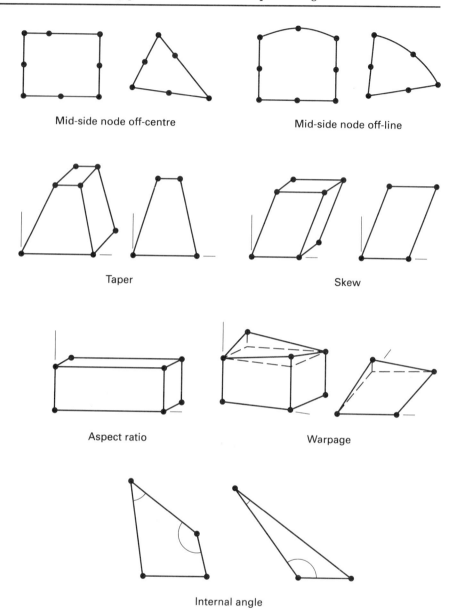

Mid-side node off-centre   Mid-side node off-line

Taper   Skew

Aspect ratio   Warpage

Internal angle

Figure 9.27 Indications of element distortion

a 'perfectly' shaped triangular element) has an aspect ratio of 1.15, while a square quadrilateral element clearly has an aspect ratio of unity. As a guide, aspect ratios of less than 3 give accurate results, while ratios between 3 and 5 are generally acceptable, but aspect ratios greater than 5 should not be used unless they occur in non-critical regions of the model.

A check on the internal angles of elements is a simple facility for finite element packages to provide, and is a very useful indicator of element

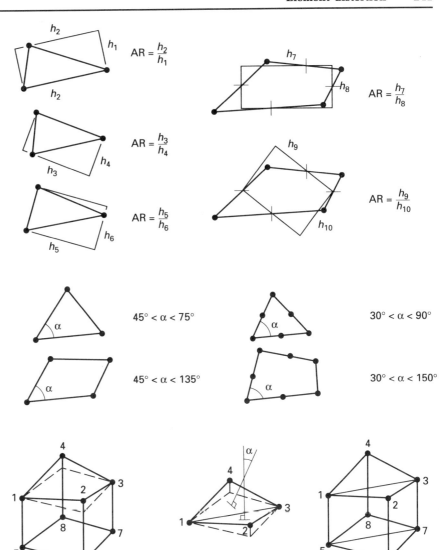

Figure 9.28 Methods of calculating the aspect ratio (AR) of triangles and quadrilaterals

Figure 9.29 Guidelines for limits of internal angles of elements

Figure 9.30 Guidelines for limits of warping of element faces

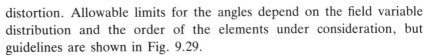

distortion. Allowable limits for the angles depend on the field variable distribution and the order of the elements under consideration, but guidelines are shown in Fig. 9.29.

Warping occurs when the nodes of a quadrilateral element (or face of a solid element) do not all lie in the same plane. As a guide, the warpage angle should not exceed 10°, measured from the normals of the two faces as illustrated in Fig. 9.30. If the warping is greater than this, then two triangular (or prism) elements should be used in place of the quadrilateral (or brick) element.

The position of the mid-side node can affect the accuracy of the solution. Ideally a mid-side node should be placed mid-way between the corner nodes. As the node moves away from the centre point and approaches the quarter position of the side, then the errors increase, until at the quarter position a singularity occurs. In practice, to minimize the likelihood of unpredictable results, the mid-side node should not be placed more than one-eighth of the side's length from the centre position, as illustrated in Fig. 9.31. Placing such a limit on the position of the mid-side node implies that the element side should not model an arc of more than 58° (Fig. 9.31). However, in areas with a high stress (or heat flow) gradient, it is advisable to model a 90° arc with at least three such (quadratic) elements.

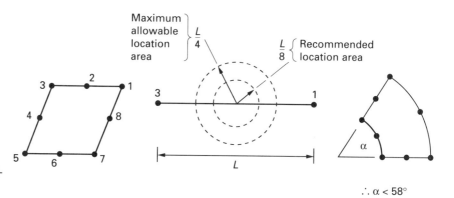

Figure 9.31 Guidelines for mid-site node location

Element distortions can be related to the Jacobian. For example, if the mid-side nodes of a complex triangular element are placed centrally between the corner nodes then the determinant is constant over all the element, but as a mid-side node moves away from the central position the Jacobian becomes variable, until a node at the quarter position causes the determinant to equal zero (see Example 8.2). Other indications of element distortion such as aspect ratio and skew can also be quantified using the Jacobian. Some commercial packages check the Jacobian at several positions in the element, usually at Gauss points and occasionally at corner nodes. A general distortion index can be devised for a particular element by calculating the ratio of the minimum Jacobian to the average Jacobian. Using such a scheme, a distortion index of 1 indicates a perfectly shaped element. Values of the index of 0.6 to 1 are generally acceptable, while ratios of less than 0.4 should be avoided.

9.5   Results processing

The finite element method solves for the nodal displacements (or temperatures), and from these the element strain and stresses (or heat flows) are calculated. The volume of output from even the smallest finite element model is large, and evaluating the performance of a component from long lists of numbers is very difficult; hence graphical presentation of the results is vital for most analyses. Examination of stress and displacement plots is also one of the quickest and easiest ways of checking that the model is constructed correctly, and is invaluable for locating any errors. The range of facilities offered by the majority of finite element packages is large. There are some sophisticated options for examining the output, for example viewing the model from any angle and viewing sections through the model. The general capabilities of post-processors are discussed in Chapter 12.

The displacements predicted by a finite element model will usually be accurate even with a coarse mesh. However, as the results of Section 9.2 show, the convergence of the displacements does not mean that the stresses have converged and are accurate. A plot of the deformed shape superimposed on the undeformed geometry is usually the first data to be examined. It gives the user an instant feel for the general behaviour of the component, while contour and vector plots of displacements give more detailed information. Displacement animation is also offered by many programs, and simulates how the component behaves under increasing load.

For simplex elements and their linear variation of displacement, a constant stress field is predicted through the element, but for complex and multiplex elements the stress varies in a linear or quadratic manner. The element matrices of these higher-order elements are calculated by numerical integration, and, after solution of the system equations, the stress is usually only calculated at the integration points. These stress values are the most accurate in the element, but not the most useful. Nodal stresses, particularly those at the corner nodes, are preferred and must be estimated by interpolating from the Gauss points, adding one further approximation into the procedure. This extrapolation of the values can lead to unreliable results at points close to local restraints or load inputs.

The stress output from an element can be referred either to its local coordinate system or to the global system. The user invariably has the choice of switching from one to another, but different elements have different default output systems. For example, plate element stresses need to be output in local coordinates to be of any use, since the local axes will lie along and perpendicular to the face of the element. The stress in the global x direction is of no use if the plate lies at 45° to the x axis. The same is obviously true for beam elements, whereas general solid elements usually require global stresses.

The use of the global or local axes for stress output has implications for the stress averaging which must be carried out on the element stresses to obtain a stress picture for the whole model. The simplest approach is to use a straightforward average of the nodal stresses whenever two or more elements connect to the same node. However, this does not take account of any size disparity of the connecting elements, and consequently some programs weight the nodal values in favour of the larger elements. The process of averaging the nodal stresses is generally acceptable at interior nodes, but is found to be less reliable at the boundary or surface nodes, and as a result these stresses should always be treated with caution.

Stress averaging across boundaries of different material properties should also be avoided, since the different materials can quite legitimately experience different stresses. Stress averaging across a boundary effectively smooths out the material interface.

The state of convergence of a model can be examined by comparing the stresses predicted by adjacent elements for a common node before stress averaging. As the model converges, so the elements should predict the same common nodal value. An example of this is shown in Fig. 9.32, which presents the nodal stresses in some of the elements of the beam bending problem discussed in Section 9.3. Note that the nodal stresses do show some variation, but the average values are reasonable. In general, if the difference in nodal values is less than 10 per cent of the maximum

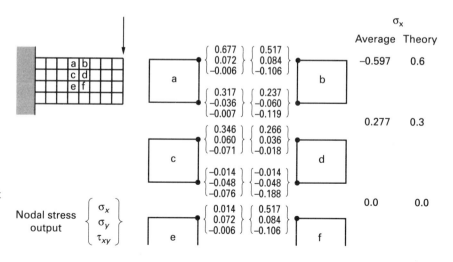

Figure 9.32 Individual element stresses for adjacent elements in a simple model, and the resultant direct stress σ_x

stress occurring in the model, the model can be considered to have converged.

For a general three-dimensional stress analysis, the basic stress output will consist of three direct and three shear stress fields, and from these can be calculated the principal stresses and the Von Mises stress. The stresses can be presented to the user in a number of ways, but contour plots are the most popular. Vector plots of principal stresses are available with some packages; these not only represent the relative magnitude of a principal stress by the length and possibly colour of the arrow, but also include the principal stress direction (Plate 4). This feature is particularly useful for examining the load flow in a component, and for displaying the effects of stress raisers in a problem.

The discussion so far has concentrated on the presentation of the results of stress analyses. Similar comments can be made about heat transfer and other field problems, although because a thermal analysis only solves for a scalar quantity, the amount of output data and necessary post-processing is markedly reduced. While simplex elements predict constant heat flows through the elements, complex and multiplex elements produce linear and quadratic variations. As with the stress calculations, the elemental heat flows are evaluated at the Gauss points and extrapolated to the nodes, where they are averaged. Graphical presentation of the results then includes contour plots of the temperature distribution and contour and vector plots of the heat flows.

9.6   Model checking

The sophistication of commercial finite element packages allows the user to construct complicated models with relative ease, and the excellent graphical output offered by the majority of the systems produces very convincing results, so that there is a great tendency to believe that the output from an analysis is always correct. However, it is all too easy to make a mistake in the development of a model or to apply the load incorrectly, producing meaningless results. For this reason, every finite element analysis should include a detailed examination of the model to confirm that it has been constructed correctly, and some verification of the results to prove that they are reasonable.

The majority of checks on the construction and loading of the model can be performed visually in the pre- and post-processors. Programs usually offer the facility of viewing from any angle and location, allowing the user to rotate the model to check that the model is constructed correctly. This can be made easier by slightly shrinking the elements, which will highlight any 'holes' in the model where elements have been inadvertently omitted, as illustrated in Fig. 9.33(b). Similarly, boundary

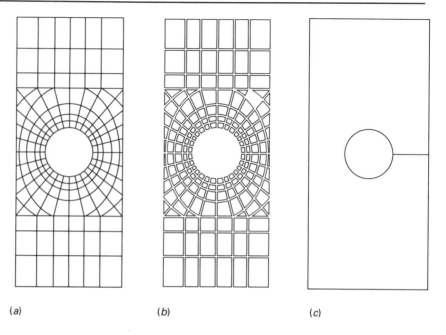

(a)   (b)   (c)

Figure 9.33 (a) Standard
element plot (b) plot with
shrink facility, showing missing
element (c) boundary line plot,
showing break in model

line plots are useful to identify regions of the model which are not
connected together in the correct way. An example of this is shown in
Fig. 9.33(c). The model appears to be constructed correctly in the standard
and shrunken element plots (Figs 9.33(a) and (b)), but the boundary line
plot shows an unexpected boundary line where part of the element is not
merged correctly. This model would execute normally, but would produce
unexpected distortion or stress plots, which ought to be detected in the
post-processing phase. Although a trivial example, this emphasizes again
how important it is for the user to understand the basic engineering of the
problem to detect any such modelling errors if they are not detected in
the pre-processing phase. Another useful facility for three-dimensional
models is the ability to produce hidden line plots (Plate 5). These can
take significantly longer to produce, but are invaluable for checking
models where the geometry and meshing are complex.

Elements which are grossly distorted should be obvious to the user, but
in any case the software usually checks for distortion and warns the user
accordingly, as discussed in Section 9.4. Depending on the user
friendliness of the program, other general checks on the input data may
also be performed; for example, unused nodes or duplicate elements may
be reported.

Many programs will plot the constraints and loads applied to the
model, allowing confirmation that they are applied in the correct location
and direction.

The final answer produced by the analysis depends on the validity and
accuracy of the model as previously discussed in this chapter, and the user
should give some consideration as to whether the model is valid and

accurate enough. The degree of validation may well depend on the previous experience of the modeller, but analyses of similar but simpler problems can give invaluable insight into the behaviour of a more complex problem. For example, the effect of assuming isotropic material properties for the orthotropic bone in the model of the natural femur in Fig. 9.2(a) could be examined by simple models using concentric cylinders rather than complex three-dimensional geometries.

Assuming that the model is a valid representation of the problem, then the accuracy of the solution should be checked where possible. The convergence of the model can be examined by h- or p-refinement techniques (see Section 9.3); these appear the most attractive methods, but are difficult to implement in practice. Consequently, it is common to compare the results with answers obtained by approximate hand calculations or by experimental observations, and one or preferably both comparisons should be performed to confirm the results of every analysis.

A detailed examination of the output from a finite element run should also be undertaken to further identify any modelling errors or inconsistencies. Defects in the model might show in the deflection or stress plots with unexpected steps or changes in gradient, but clearly for the user to recognize such unusual results he must have a sound understanding of the engineering of the problem. There are several fundamental checks that should be carried out on the output to confirm that the model is reasonable. These include:

(a) Volume and mass checks: some programs calculate the volume and/or mass of the model, and this is useful to confirm that the geometry is represented correctly.

(b) Reaction force checks: the sum of all the reaction forces should equal the applied forces.

(c) Model equilibrium: the sum of the forces and resultant moments acting on the problem should equal zero.

(d) The results should be consistent with the problem formulation. For example, for a linear static analysis the displacements should be 'small', and the materials should not exceed their elastic limits.

(e) The results should be symmetric for problems with symmetry.

(f) Some software confirms that the derived displacements multiplied by the stiffness matrix do indeed equal the applied forces (within limits).

It is worth noting that the majority of programs automatically scale displacement plots before they are plotted on the screen; thus incorrectly constrained models producing excessive displacements can appear reasonable at a first glance. Not only the form but also the magnitude of the displacements should be checked.

Before the system equations are solved, most commercial programs make some checks on the condition of the stiffness matrix, to confirm that it is suitable. The two most likely problems are singularities and ill-conditioning of the stiffness matrix.

Ill-conditioning occurs when there is a large difference in the magnitude of the terms in the stiffness matrix, making the solution susceptible to round-off errors. Programs usually check for ill-conditioning by comparing the largest and smallest terms in the stiffness matrix. If the ratio of the magnitudes typically exceeds 1×10^8, a warning is issued. There are several possible causes of ill-conditioning, but the most common is a large difference in the stiffness of parts of the model. This may be as a result of large differences in element size and/or material properties in a particular model, or even large stiffness differences in a single element. A common example of this latter condition occurs when thin shells are modelled with solid elements. Such thin shells should be modelled with special thin shell elements (which ignore transverse shear effects). Ill-conditioning leads to an overall decrease in the accuracy of the solution, and can usually be detected by simple reaction force checks. It is generally not a problem with very accurate solution procedures.

Singularities occur when an indeterminate or non-unique solution is possible. If a singularity is detected during a solution the analysis may abort, or possibly continue with a warning message advising of negative or zero main diagonal terms; if the analysis continues, the resulting values of field variable will be excessively high. Singularities can be caused by a number of conditions, but the most common are:

(a) An unconstrained structure, where the complete model or any part connected through non-linear elements (such as gap elements) is not adequately constrained. All models used to represent static problems must be adequately constrained, so that both rotational and translational rigid body motions are prevented.

(b) An unconstrained joint. For example, two collinear horizontal spar elements will have an unconstrained vertical degree of freedom at the joint.

(c) An incomplete specification of real constants. For example, omission of some of the thicknesses at the nodes of shell or plane stress elements will lead to a singularity.

(d) Incorrect material properties. Obvious examples of this are zero elastic modulus or thermal conductivity, and a Poisson's ratio equal to 0.5.

Owing to the limited word size on all computers, some singularities may appear as ill-conditioning of the matrix, allowing the solution of the equations. However, such ill-conditioning results in large rigid body motions with relatively accurate reaction force checks.

10 Further applications of the finite element method

10.1 Introduction

The previous chapters of this book have considered linear static and steady state applications, but the finite element method can also be used to analyse non-linear static, dynamic and transient problems. The range of possible applications is vast and impossible to cover in detail in a single book, but this chapter briefly introduces the types of problem that can be analysed, and discusses some of the solution techniques. Examples of the range of problems that can be analysed are given in Chapter 1, but of necessity this chapter concentrates only on the areas of solid mechanics and heat transfer.

The application of the finite element method to the solution of non-linear static, dynamic or transient problems is certainly more complex than that to the linear and steady state problems discussed so far, and the use of commercial finite element software for the analysis of such problems usually requires specialized training. Because the problems are non-linear or time dependent, their solutions invariably require more than a single step. They may involve a load step or time step increment, but fortunately most commercial programs include routines to optimize the stepping, so that the solution can be reached in the minimum number of iterations. Notwithstanding these facilities, the types of problem introduced below usually demand a significant increase in computer time and resources.

10.2 Non-linear static elasticity problems

Non-linear static problems are those where the stiffness matrix $[k]$ and/or the force vector $\{F\}$ are functions of the nodal displacements $\{U\}$. This may occur as a result of non-linearities in the material properties, the geometry, combined effects (material and geometry), or the contact conditions of the problem.

(*a*)

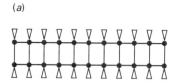

(*b*)   Radial stress (MPa)   Axial stress (MPa)

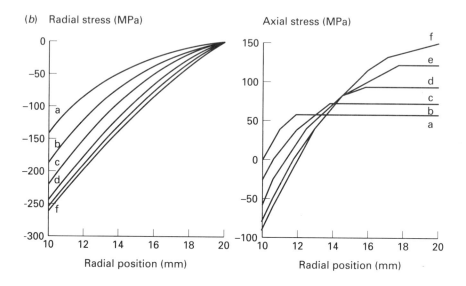

Line	Load (% of maximum)
a	55
b	72
c	85
d	94
e	98
f	100

a = all elastic
f = all plastic

Hoop stress (MPa)

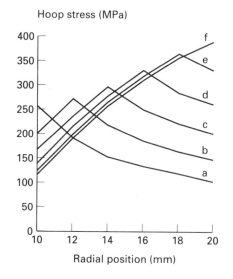

Figure 10.1 Elastic-plastic analysis of a thick cylinder: (a) finite element model of thick cylinder with fixed ends (i.e. no axial strain), using axisymmetric elements (b) stress plots

10.2.1 Material non-linearity

A linear static analysis assumes that the material is within its elastic limit, and that it follows a simple linear stress–strain curve. Problems where this is not the case include those exhibiting plasticity and creep of the material. An example of an elastic-plastic analysis is shown in Fig. 10.1. The problem is a cylinder under an internal pressure loading which is sufficient to cause the complete thickness to become plastic. Plots of the stress distribution are presented for increasing pressures, showing the development of the plastic yielding through the cylinder wall thickness.

For such plasticity problems an idealized stress–strain curve must be supplied to the finite element program, and is usually approximated in a bilinear or multilinear way, depending on the particular material, as shown in Fig. 10.2. The elastic-plastic behaviour of the material is derived from various yield criteria and hardening rules. For example, yielding might be based on the Von Mises criteria, with the plastic strain increments calculated using the Prandtl-Reuss flow rule, and the plastic deformation described by either an isotropic or a kinematic hardening model. The loading is applied incrementally to the model, and within each load step an iterative solution is performed until the stress–strain values at the Gauss points are correct.

Creep occurs when the loading is applied over an extended period causing a permanent deformation, even though the induced stresses are below the yield stress of the material. The form of a typical creep curve is shown in Fig. 10.3. It is divided into four regions, namely an elastic section governed by Hooke's law, and then the primary, secondary and tertiary creep stages. The tertiary stage is usually neglected because the component is close to failure. Primary and secondary stage creep are governed by equations of the following forms:

Figure 10.2 Idealized stress–strain curves for plasticity analyses

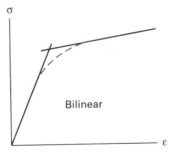
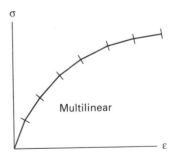

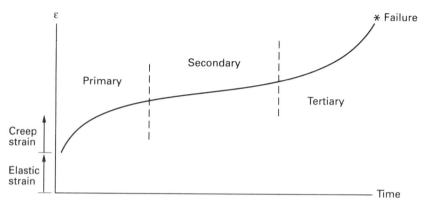

Figure 10.3 Typical creep curve

$$\text{primary creep} \qquad \Delta\varepsilon^{cr} = c_1\sigma^{c_2}t^{c_3}e^{-c_4/T}\Delta t$$
$$\text{secondary creep} \qquad \Delta\varepsilon^{cr} = c_5\sigma^{c_6}e^{-c_7/T}\Delta t \qquad\qquad [10.1]$$

where c_1 to c_7 are constants, σ is the equivalent stress, t is the time at the end of the iteration, and T is temperature.

Creep problems are solved by an incremental technique similar to that used for plasticity. Figure 10.4 shows the creep of a cylinder similar to that in Fig. 10.1 when subjected to a constant internal pressure. Note that the internal and external radii increase as expected, but also that the internal stress distributions are quite different from the static solution.

Some finite element programs also have the capability to model swelling, which is the volumetric enlargement of the material due to neutron bombardment or other effects. The swelling strain rate might be a function of the temperature, time, neutron flux level or stress.

Hyperelastic materials such as rubber have linear properties but can experience large elastic strains. Consequently they require a large-displacement analysis and are included in the following section.

10.2.2 Geometric non-linearity

A large-displacement analysis is required when the structure's displacements become so large that the original stiffness matrix no longer adequately represents the structure. Large-displacement problems may be divided into two areas: those which result in small (element) strains, and those resulting in large finite strains. The small-strain condition implies that the material remains elastic, and consequently that the structure returns to its original configuration when the loading is removed. An example of a large-deflection (small-strain) analysis of a cantilever beam is shown in Plate 6. Those elements experiencing large strains undergo permanent deformations, with the exception of hyperelastic materials

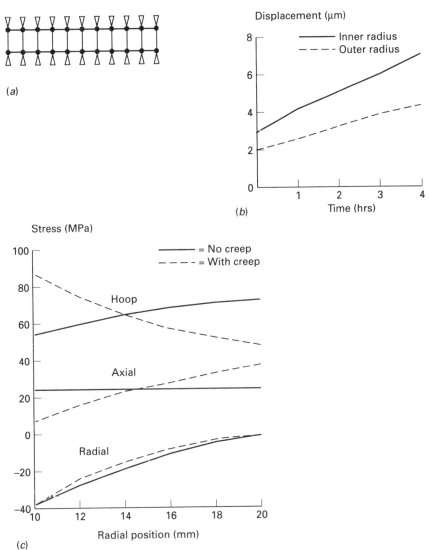

Figure 10.4 Creep analysis of a thick cylinder: (a) finite element model of thick cylinder with fixed ends (i.e. no axial strain), using axisymmetric elements (b) radial deformation of the inner and outer radii with increasing time (c) comparison of stress distributions through the cylinder for models with and without creep

(such as rubber) which can elastically accommodate high strains. The permanent material deformation requires non-linear material properties as discussed in the previous section, and thus this class of problem has combined material and geometric non-linearities. Such problems are examined in Section 10.2.3.

In large-displacement problems, care needs to be taken with the applied loading as the geometry changes, because a nodal load will not generally rotate with the node, while surface pressures will follow the model as it moves (Fig. 10.5). Body forces continue to act in their original directions. To assist the user some programs include the option of

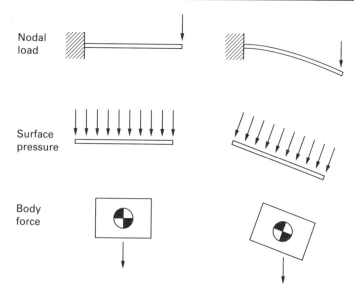

Figure 10.5 Load directions
during large-deflection analyses

defining follower or non-follower nodal loads, where the former is automatically reoriented during the analysis to maintain its position relative to the node.

If a structure undergoes a large displacement due to an applied load then, since its geometry changes, its stiffness matrix needs to be adjusted accordingly. There are two ways in which this can be achieved. The first approximate method assumes that the size of the individual elements is constant, so that a reorientation of the elemental stiffness matrices due to the elements' rotation and/or translation is all that is required. The second method is more accurate, and recalculates the stiffness matrices of the elements after adjusting the nodal coordinates with the calculated displacements. In both cases an incremental solution is performed, usually by the Newton-Raphson method. If the spatial motions are not large then it is possible to apply the load in a single step with several iterations, but for large motions the load is best applied in smaller steps.

Materials with hyperelastic properties require the nodal coordinates to be updated during their loading. Rubber is the most common example of such a material, but certain plastics can also undergo very large elastic strains. Plate 7 shows the deflection of a rubber sheet with a central hole.

Stress stiffening (or geometric stiffening) is also a geometric non-linearity; it is the stiffening (or weakening) of a structure due to the state of stress. The system equations for such problems are

$$([k] + [k_G])\{U\} = \{F\} \qquad\qquad [10.2]$$

where $[k_G]$ is the geometric (or initial stress) stiffness matrix and is a function of the state of stress in the structure. The equation is clearly non-linear, because the state of stress depends on the displacements. The equations are solved iteratively. In the first iteration $[k_G]$ is ignored and a normal static analysis is performed, and in each subsequent iteration $[k_G]$ is calculated from the state of stress of the previous iteration. (Convergence usually requires fewer than three iterations.)

Stress stiffening is important in structures weak in bending resistance, held in place by tensile (or compressive) stresses. A trivial example is a horizontal strut, pin-jointed at one end and simply supported at the other, subjected to a horizontal 'tensioning' force, as shown in Fig. 10.6. If the strut is loaded by a vertical force at its mid-point, then its lateral stiffness is dependent on the horizontal force. The stiffness is increased by the presence of a positive tensioning force, but decreased if the horizontal force is compressive. Other examples where stress stiffening is important include membranes and cables.

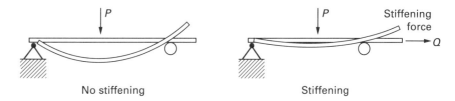

Figure 10.6 Example where stress stiffening is important

No stiffening   Stiffening

It is quite conceivable that large-displacement problems can themselves experience stress stiffening effects, in which case the geometric stiffness matrix must also be included in the large-displacement solution.

10.2.3 Combined material and geometric non-linearity

Problems requiring both material and geometric non-linearities are probably the most demanding type of analyses that can be undertaken by the finite element method, and include metal working processes such as forming, rolling and extrusion. An example of this type of problem is shown in Plate 8.

One of the most difficult parts of modelling large-deformation analyses is the implementation of the boundary conditions, which can include for example various contact conditions and sliding along curved surfaces with and without friction. The facility to model non-linear boundary conditions with non-linear materials and geometries is found in only a few very specialized finite element programs, which require considerable training and expertise to use effectively and reliably.

10.2.4 Non-linear contact conditions

The finite element method assumes a perfect connection between two parts of a model unless special non-linear elements are used. There are two types of non-linear contact elements most commonly available, namely tension-only and gap (or compression-only) elements. For example, a short cable could be modelled by a tension-only element. Gap elements were briefly introduced in Section 9.2 to represent the contact between a pin and a lug. Usually finite element programs contain two- and three-dimensional versions of gap elements. The elements can transmit a frictional force as well as a compressive force, and some elements may also permit relative sliding of the two contacting surfaces. The initial conditions may be specified as either an initial gap between the two parts of the model or a preload, implying that the two surfaces are pushed against each other.

All non-linear contact conditions require an iterative solution. After each iteration, the state of every element is examined and adjustments are made as necessary. For example, a compression-only spar is disconnected if it is found to be in tension. The model is assumed to have converged when the condition of every element remains unchanged.

10.3   Buckling problems

Buckling problems are concerned with the calculation of the critical loads to cause elastic instability of a structure. The most frequently encountered example is buckling of a strut, where an increasing axial load suddenly causes a critical failure. Other examples include the buckling of plates and externally pressurized cylinders.

A theoretical analysis of buckling problems shows that they are eigenvalue problems, governed by the following equation:

$$([k] + \lambda[k_G])\{U\} = 0 \tag{10.3}$$

where $[k_G]$ is the geometric stiffness matrix, previously encountered with stress stiffening problems; λ is the eigenvalue, related to the buckling load; and $\{U\}$ is the associated vector of nodal displacements describing the mode shape. According to the theory, there will be a number of buckling loads and mode shapes. Plate 9 shows the first four buckling modes of a flat square plate subjected to in-plane edge loads.

Unfortunately, however, this method does not take account of any initial imperfections in the structure or component, and the results rarely agree with those determined in practice. For the square plate, the buckling loads will not be correct unless the plate is perfectly flat, and the loads are applied exactly in the plane of the plate. Therefore, instead of

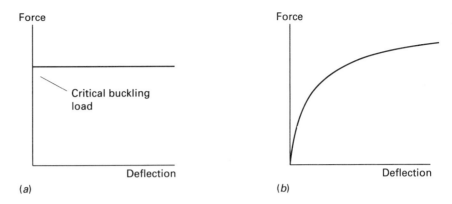

Force   Force

Critical buckling
load

Figure 10.7 Load–deflection
graphs defining buckling: (a)
idealized behaviour (b) actual
behaviour

Deflection   Deflection

(*a*)   (*b*)

the load–deflection graph being similar to that illustrated in Fig. 10.7(a),
it is more likely to be of the type shown in Fig. 10.7(b). Thus a sudden
buckling is unlikely to occur. Instead of using Eq. 10.3, buckling
problems should be solved by a large-displacement static analysis,
possibly with stress stiffening as outlined in Section 10.2.2. Using a large-
displacement analysis, buckling can be detected by the change of
displacement at any node. If the displacement change is increasing, the
load is above critical and buckling is occurring. If the change is
decreasing, the load is below the buckling load.

   In summary, then, the eigenvalue method of solution of buckling is
usually approximate, and overestimates the buckling load, *i.e.* produces
unsafe answers. It also gives no information about the post-buckling
behaviour. A large-displacement analysis, on the other hand, is accurate
and does give the complete behaviour of the component before and after
buckling.

10.4   Dynamic problems

In a dynamic analysis the effects of inertia forces are included in the
calculations. These inertia forces are proportional to the acceleration of
the body under investigation, and as a result introduce a time variation
into the system equations. Solution of the equations yields some form of
time varying or dynamic response of the body.

   The basic equations for the dynamic behaviour of a structure or
component can be expressed as

$$[M]\{\ddot{U}\} + [C]\{\dot{U}\} + [k]\{U\} = \{F(t)\}$$ [10.4]

where $[M]$ is the total mass matrix of the structure, found by summing the
individual element mass matrices; $[C]$ is the structural damping matrix;
$[k]$ is the stiffness matrix; $\{\ddot{U}\}$, $\{\dot{U}\}$, $\{U\}$ are the nodal accelerations,

velocities and displacements respectively; and $\{F(t)\}$ is the vector of applied forces, which may be functions of time t.

The finite element method may be used to analyse a wide range of dynamic problems, and four of the main application areas are discussed below.

10.4.1 Modal analysis

This is concerned with the prediction of the natural frequencies and mode shapes of undamped structures under free vibrations, and is the most frequently performed type of dynamic analysis. It is important that the engineer knows the natural frequencies of certain bodies to ensure that they are not excited by any applied loading, which would result in high-amplitude vibrations. An obvious example where these effects might be critical is in helicopter design. If any components in the helicopter have natural frequencies that are close to the rotational speed of the rotors, then resonance of a component could occur, leading for example to a possible fatigue failure. Other examples of such eigenvalue problems are given in Table 1.2. An example of the results of such an analysis is shown in Plate 10, which shows the first mode shape of a pump casing.

For eigenvalue analyses, the governing Eq. 10.4 reduces to

$$[M]\{\ddot{U}\} + [k]\{U\} = 0 \tag{10.5}$$

If the vibration is assumed to be sinusoidal in form, then

$$u = A\cos(\omega t) \quad \text{and} \quad \ddot{u} = -\omega^2 A\cos(\omega t) = -\omega^2 u \tag{10.6}$$

so that Eq. 10.5 becomes

$$([k] - \omega^2[M])\{U\} = 0 \tag{10.7}$$

In this equation both ω^2 and $\{U\}$ are unknown. ω^2 is known as the eigenvalue, while $\{U\}$ is the normalized eigenvector and describes the shape of the structure as it vibrates. If a model has n degrees of freedom, then there will be n possible different combinations of ω^2 and $\{U\}$; in other words, the body will have n natural frequencies and associated mode shapes. Fortunately, however, the engineer is usually only interested in a limited number of the lowest frequencies. An example of a range of mode shapes is presented in Plate 11, which shows the first four natural frequencies and mode shapes of a square plate that is simply supported on two sides.

Modal analyses are generally much more costly than static analyses. For example, consider a modal and a static analysis of a problem with 400 degrees of freedom and a wavefront of 30. It might take of the order of five times the number of operations to calculate one eigenvalue compared with one static load case, or fifteen times for four eigenvalues compared with four static load cases. It usually becomes impracticable therefore to solve dynamic equations with all the system unknowns; some form of simplification is required so that only the degrees of freedom necessary to characterize the behaviour of the system are considered. This process is known as static condensation, and involves the selection of 'master' and 'slave' degrees of freedom. The master degrees of freedom are chosen so that they define the lowest modes and frequencies of the structure, while the slaves are related to the highest modes. For example, for a beam problem where bending is of primary importance, the rotational and axial (stretching) degrees of freedom can be chosen as slaves. It is possible to formularize the selection procedure and many programs automatically select master degrees of freedom for the user. Once the master degrees of freedom are identified, the slaves are condensed out of the system equations using a technique known as Guyan reduction. The result of the static condensation is a reduced set of system equations which quite adequately describe the dynamic behaviour of the system, and which can be solved economically.

There are a number of methods available for the solution of the equations. The best method will vary from one problem to another, and depends on the size and form of the equations being solved and the number of eigenvalues and eigenvectors that are required. The most popular methods include the generalized Jacobian, Householder, Givens, Lanczos and subspace iteration.

10.4.2 Transient response analysis

A transient response analysis determines the response of a structure due to time varying loads, which can be forces that vary with time, or alternatively a time function of displacement, velocity or acceleration. The governing equation for these problems is given by Eq. 10.4. A simple analysis of a beam, built in at both ends, is illustrated in Fig. 10.8. The load is increased linearly up to a maximum, held at that value for a short time and then decreased to a constant value. The resulting displacement, velocity and acceleration of the centre of the beam are presented for 0 to 0.25 seconds after the load is first applied. (Note that since there is no damping assumed in this model, the output ultimately settles down to the natural frequency of the beam.)

Transient problems may be solved in two fundamentally different ways: either integration over the time domain, or solution in the frequency domain. In the first of these, the forcing function is divided into a number of impulses which can be integrated over time. In the second, the forcing function is decomposed into its frequency components and the solution is

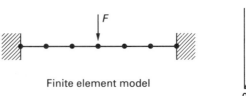

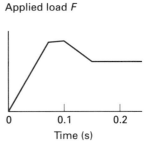

Applied load *F*

Finite element model

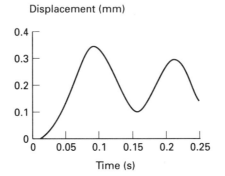

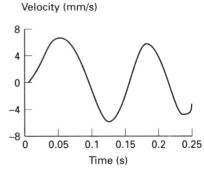

Displacement (mm)

Velocity (mm/s)

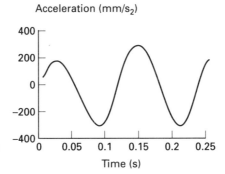

Acceleration (mm/s$_2$)

Figure 10.8 Transient analysis of a built-in beam subjected to a variable load

found in the frequency domain. This latter (fast Fourier transform) approach is not widely used.

There are two different methods of time domain solution, namely direct integration and modal superposition. Direct integration is the most general method, and can be used for both linear and non-linear problems. On the other hand, modal superposition is a linear superposition, and consequently is only suitable for linear systems, that is those systems where $[k]$ and $[M]$ are constant. Furthermore, the modal superposition method cannot take account of time varying imposed displacements.

The direct integration method involves a step-by-step solution of the equations of motion. The most popular integration scheme is the Newmark-β method, which is implicit and unconditionally stable. An implicit integration method is one where the equations are solved at time $t + \mathrm{d}t$ in terms of the variables at time $t + \mathrm{d}t$, and therefore requires an inversion of the system equations. Unconditional stability means that any errors will not grow, irrespective of the time step. The following approximations are made:

$$\{\dot{U}\}_{t+\mathrm{d}t} = \{\dot{U}\}_t + (\{\ddot{U}\}_t + \{\ddot{U}\}_{t+\mathrm{d}t})\mathrm{d}t/2 \qquad [10.8]$$

$$\{U\}_{t+\mathrm{d}t} = \{U\}_t + \{\dot{U}\}_t\mathrm{d}t + ((\tfrac{1}{2} - \beta)\{\ddot{U}\}_t + \beta\{\ddot{U}\}_{t+\mathrm{d}t})(\mathrm{d}t)^2 \quad [10.9]$$

where β depends on an amplitude decay factor, but usually a value of 0.25 is used.

Analysis by modal superposition involves firstly the calculation of the natural frequencies of the problem, as outlined in the previous section, and then the combination of the different ratios of the modal responses to predict the time varying behaviour. Briefly, the method is as follows. The variation of the displacements with time is found from

$$\{U\} = \sum_{i=1}^{n} Y_i\{v\}_i \qquad [10.10]$$

where n is the number of modes, $\{v\}_i$ is the mode shape of mode i and Y_i is a general modal coordinate which is a function of time, and satisfies

$$\ddot{Y}_i + 2\xi_i\omega_iY_i + \omega_i^2Y_i = P_i = \{v\}_i^{\mathrm{T}}\{F(t)\} \qquad [10.11]$$

with ω_i being the frequency of mode i and ξ_i the modal damping ratio. If there are n modes under consideration, there will be n such equations, which are solved by the Newark direct integration method discussed previously. Once the variations of the n generalized coordinates are determined, the transient response of the structure can be calculated from Eq. 10.10. Figure 10.9 shows the addition of the first three modes of a problem.

The choice between direct integration or modal superposition depends on the nature of the loading and the type of response required. The major cost with modal superposition is in the calculation of the modes, and the number of the modes required depends on the highest-frequency component excited by the load. For example, a sudden shock load will excite many modes, and for a short-response analysis a modal superposition would be more expensive than direct integration. However, if a long response is required or if several load cases are to be analysed, the first method is preferred. Where the loading only excites a few modes, as in the case of an earthquake, modal superposition is the most obvious choice. In most cases no more than ten of the lowest modes are

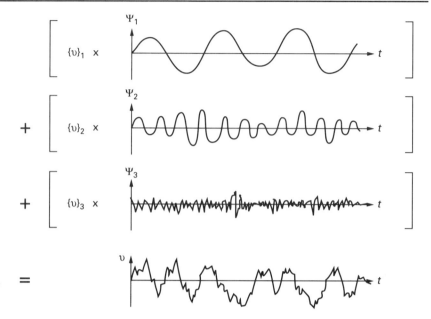

Figure 10.9 Modal superposition for the analysis of a transient problem

required for a transient analysis.

As with the solution of the modal problem in Section 10.4.1, it is not usually necessary to use all the degrees of freedom in a transient analysis. A reduced, linear transient analysis is usually possible with only a limited number of master degrees of freedom. Where such a reduced analysis is chosen, the first stage of the calculations produces the displacement time history of the master degrees of freedom (sometimes referred to as the 'displacement pass'), and a second stage (or 'stress pass') then calculates the displacements of the slave degrees of freedom and the element stresses.

A transient analysis can produce predictions of nodal displacements, velocities and accelerations, together with element stresses at every time step. Indeed the sheer volume of data produced during a transient analysis can be a significant problem. Since only a small fraction of the output is usually of interest, it is vital that careful consideration is given to precisely what is required, and the output routines controlled accordingly.

10.4.3 Harmonic response analysis

Harmonic response analysis is used to find the steady state response of a structure to a set of harmonic loads of known amplitude and frequency. (A harmonic load is one that varies sinusoidally in time at a single frequency; see Fig. 10.10.) Note, however, that even though the loads must be at the same frequency, phase and amplitude differences between

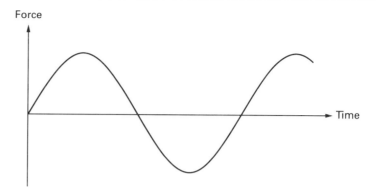

Figure 10.10 Harmonic load form

the loads may be specified at different degrees of freedom. Harmonic loads are usually man-made, and occur for example in rotating or reciprocating machinery. The response of the structure or component under investigation will depend on its natural frequencies and the frequency of the applied loading, but the harmonic response analysis will produce predictions of the magnitude and phase of the nodal displacements, velocities and accelerations, together with element stresses and forces. If the response of the structure is required at various frequencies, then the process can be repeated over a range of frequency values, resulting in graphs of frequency versus nodal displacement, velocity, acceleration or stress. Some finite element packages can perform such 'frequency sweeps' automatically.

An example of the type of output that the method can produce is shown in Fig. 10.11. The results show the variation of the horizontal displacement at a particular node when the applied loading has a frequency of f Hz, and compares the maximum displacement at the same node with that predicted for a range of other operating frequencies. The results give a good estimate of the natural frequencies of the model, f_1 and f_2.

Two methods are commonly used for the solution of harmonic analysis problems: a direct method, and a modal superposition method. As the name suggests, the first method solves the system equations directly, as follows. If the loads are sinusoidal with frequency ω, then the displacements will also vary with the same frequency but with a phase lag ϕ. For example,

$$F(t) = F_0 \cos\omega t$$

Hence

$$u = A\cos(\omega t - \phi) \quad \text{and} \quad \ddot{u} = -\omega^2 A\cos(\omega t - \phi) = -\omega^2 u$$

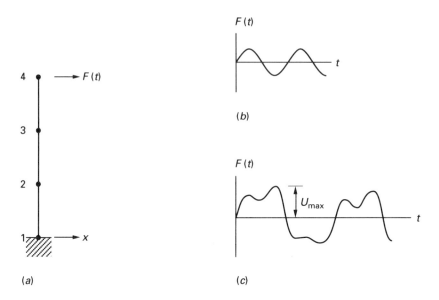

(a)

(b)

(c)

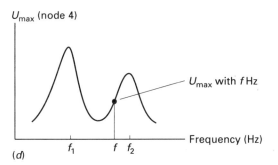

Figure 10.11 Sample output data from a harmonic response analysis: (a) finite element model, (b) applied load, (c) variation of the horizontal displacement u_{max} at node 4, when the forcing frequency is f Hz, (d) variation of u_{max} of node 4 with different forcing frequencies

Therefore the general system equation becomes

$$([k] - \omega^2[M])\{U\} = \{F_0\cos\omega t\} \qquad [10.12]$$

Since F_0 and ω are known, Eq. 10.12 can be solved directly for $\{U\}$.

Analysis of harmonic response problems by modal superposition is similar to the method already described in Section 10.4.2 for transient problems. However, the general modal coordinate Y_i in Eqs 10.10 and

10.11 can be shown to be a simple function of the forcing frequency, *i.e.*

$$Y_i = \left(\frac{P_i}{\omega_i^2} \right) \beta_i \cos(\omega t - \phi_i) \tag{10.13}$$

β_i is known as the amplification factor for mode i, and both β_i and ϕ_i are simple functions of ω, ω_i and ξ_i. Therefore, having calculated Y_i, the nodal displacements are simply calculated from Eq. 10.10 in the usual way. Modal superposition is only suitable for harmonic analysis of linear systems.

It is usually possible to use reduced mass and stiffness matrices in harmonic analysis by the selection of master degrees of freedom as described previously. After the displacements of the master degrees of freedom are calculated, a stress pass is required to derive the displacements of the slaves together with the element stresses.

10.4.4 Shock spectrum analysis

Shock spectrum analysis, sometimes referred to as response spectrum analysis, can be used to estimate the maximum response of a structure to an arbitrary foundation shock load. The type of loading that is considered is shown in Fig. 10.12. The complexity of the loading means that normal time marching methods cannot be used, and response spectra must be considered instead. The solution of these problems requires three steps. Firstly, a single degree of freedom system (Fig. 10.13) is analysed under the defined shock load. The governing equation for such a system is most conveniently expressed in a format different to that used in Eq. 10.4, namely

$$\ddot{v} + 2\xi\omega\dot{v} + \omega^2 v = \ddot{v}_g \tag{10.14}$$

where $\omega = \sqrt{(k/M)}$ is the natural frequency (rad/s), $\xi = C/2M\omega$ is the damping ratio, $\ddot{v}_g$ is the ground shock acceleration (Fig. 10.13), C is the damping constant, M is the mass and k is the stiffness.

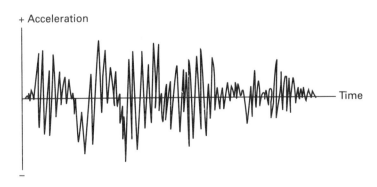

Figure 10.12 Earthquake accelerations

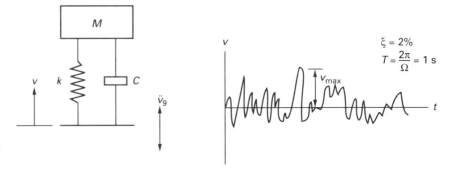

Figure 10.13 Response of a single degree of freedom to the shock load

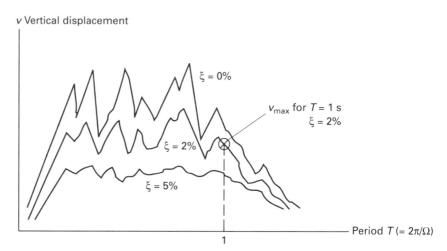

Figure 10.14 Vertical displacement response spectra

The analysis predicts the variation of the vertical displacement, velocity or acceleration with time as illustrated by the sample displacement graph in Fig. 10.13. The result of interest is the maximum value, in this case of the displacement, v_{max}.

If the period $(=2\pi/\omega)$ and damping ratio of the system are varied, and the same shock load history is applied, then the results can be combined to produce graphs of maximum vertical displacement as a function of period and damping ratio, as illustrated in Fig. 10.14. These are the displacement response spectra for the specified ground acceleration. In some cases such response spectra may be 'standardized' and supplied as design spectra, rather than the actual ground acceleration being specified. For example, standard design spectra are laid down for the design of nuclear power stations.

The second stage of the procedure is to perform a modal analysis of the real structure to determine its natural frequencies and mode shapes (see Section 10.4.1). Finally, the mode shape displacements (which are normalized so that the maximum displacement is one unit) are factored by the corresponding magnitude obtained from the displacement spectra to give the real structural displacement. Combination of the responses of all the relevant modes then gives the total response of the structure to the shock loading. A number of combination schemes are usually available, such as the square root of the sum of the squares (SRSS) and the complete quadratic combination (CQC).

10.5   Transient thermal problems

The basic equation of transient thermal problems is

$$[C]\{\dot{\Phi}\} + [k]\{\Phi\} = \{F\} \qquad [10.15]$$

where $[C]$ is the specific heat matrix and $[k]$ is the thermal conductivity matrix. The equation is similar to that used for dynamic structural problems (Eq. 10.4), except that there is no 'inertia' term. Transient thermal problems are easier to solve than dynamic problems because the order of the thermal equation is one order less, and temperature is a scalar whereas structural analyses have three (or possibly six) degrees of freedom to solve for at each node.

Several methods are currently used in the solution of these transient problems, including Crank-Nicholson and a method based on the Houbolt approach. The Crank-Nicholson method is a central difference scheme in which the temperature at time $t + dt$ is evaluated from the value at time t. The Houbolt-based method assumes a quadratic variation of the temperature with time, and the value at time $t + dt$ is calculated from the temperature at times t and $t - dt$. The integration time step dt can vary during these types of analyses, and many programs include optimization procedures which adjust the time step automatically.

An example of a transient thermal analysis is shown in Plate 12. It shows the temperature distribution in a pipe with a supporting flange when a liquid at 200 °C is suddenly flushed along it.

11 Symmetry, submodelling and validation

11.1 Introduction

The different types of symmetry that a modeller can expect to encounter in a problem are discussed in Chapter 3. By careful model design it is shown that advantage can be taken of this symmetry, and only a portion of the actual problem needs to be modelled. It might appear at first sight that symmetry can only be considered if the geometry, displacement boundary conditions and loads are symmetric. However, the geometry and boundary conditions must certainly be symmetric, but unsymmetric loading can be considered in a symmetric model. This occurs most frequently in planar and axial symmetry, and many commercial packages include special facilities to aid in the analysis of such problems. These two types of symmetry are discussed in detail in this chapter.

The techniques of submodelling and substructuring are also introduced. These are two very useful techniques for dealing with large and/or complex problems, and are particularly suitable where just one part of the structure needs to be analysed repeatedly.

The concepts of the validity and the accuracy of finite element models are presented in Chapter 9, and relate to the representation of the physical problem and the accuracy of the consequent solution respectively. They depend on the correct application of the finite element software, but assume that the numerical routines are implemented correctly. Methods of checking the accuracy of the computer implementation of the finite element procedures are discussed in this chapter, and are concerned with three types of element and program validation tests.

11.2 Symmetric models with non-symmetric loading

Planar symmetry

Any arbitrary loading on a symmetric structure can be decomposed into symmetric and antisymmetric components, as illustrated by the simple (single plane of symmetry) examples in Fig. 11.1. (Antisymmetric loading components are ones where the load profiles and points of application are symmetric, but the load directions are not symmetric.) Understandably, an analysis will only predict the right answers if the correct boundary conditions are applied to all the nodes on the plane(s) of symmetry. For models with symmetric loading these boundary conditions are usually easy to derive, but for the antisymmetric model they are not so obvious. Fortunately, however, there is a simple relationship between the two cases. Consider for example the unsymmetric loading of the framework in Fig. 11.1(a). The expected displacement of the two loading components, together with the two finite element models required for the analysis, are shown in Fig. 11.2. The boundary conditions for the symmetry plane of the two models are also included. (The half framework is represented by three beam elements.) This particular problem is only two-dimensional, and the beam element has three degrees of freedom at each node (namely two translations, u and v in the x and y directions respectively, and one rotation θ_z about the z axis; see Section 8.5). The degrees of freedom of the node on the centre-line dictate whether symmetry or antisymmetry is being modelled. The degrees of freedom may be fixed (*i.e.* zero) or free, and are as follows:

	u	v	θ_z
Symmetry	Fixed	Free	Fixed
Antisymmetry	Free	Fixed	Free

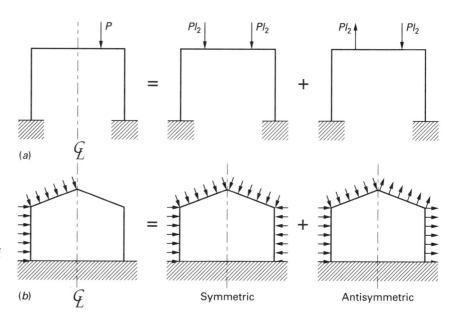

Figure 11.1 Decomposition of unsymmetric loading into symmetric and antisymmetric components

The symmetry and antisymmetry conditions are exactly opposite to each other. With two and three planes of symmetry, the number of load component combinations increases to four and eight respectively, and the relationship between the symmetry and antisymmetry boundary conditions can prove most useful. Some programs will automatically apply the correct constraints for both symmetric and antisymmetric boundaries.

After the load components have been analysed individually, the results are combined. For the problem in Fig. 11.2 this involves summing the answers to obtain the results for the half of the problem that was analysed, and subtracting the two sets of values for the part that was not modelled. For multiple planes of symmetry, this post-processing can become complicated and easily susceptible to human error, and consequently some commercial programs include automated post-processing facilities.

Although multiple runs of symmetry problems with non-symmetric loading are required, the analysis costs are generally considerably less than the analysis of the whole structure. The pre- and post-processing does, however, require considerably more care to ensure that the application of the boundary conditions and the combination of the results are performed correctly.

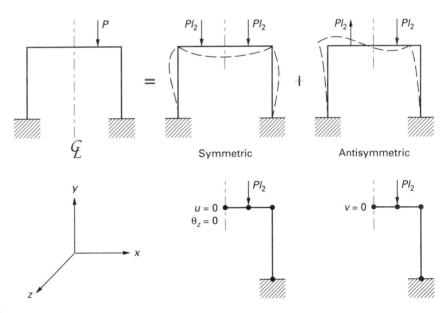

Figure 11.2 Finite element models and boundary conditions of the symmetric and antisymmetric load components

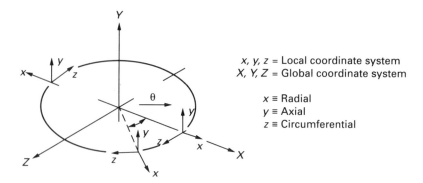

Figure 11.3 The local coordinate system in relation to the global system

Axial symmetry

The application of non-symmetric loads to axisymmetric problems requires a different approach to that used for planar symmetry. For this situation, the non-axisymmetric loading is expressed as the sum of several component loadings (or harmonics) by the Fourier series method. Each harmonic load is applied separately to a two-dimensional (axisymmetric) model, and the solution to the original problem is then calculated by combining the responses of the individual loads.

The Fourier series of a load in terms of the position θ around the circumference is

$$F(\theta) = \sum_{n=0}^{\infty} F_n \cos(n\theta) + \sum_{n=0}^{\infty} F'_n \sin(n\theta) \qquad [11.1]$$

where F_n and F'_n are the magnitudes of the forces. Usually when axisymmetric models are generated, they are defined in the x–y plane, where the x axis is in the radial direction, the y axis lies along the axis of symmetry and the z axis is directed circumferentially. The angle θ then varies as illustrated in Fig. 11.3.

The forces in each of the Cartesian directions can be expressed in the form of Eq. 11.1 as follows:

$$F_x = \sum_{n=0}^{\infty} F_{xn} \cos(n\theta) + \sum_{n=0}^{\infty} F'_{xn} \sin(n\theta)$$

$$F_y = \sum_{n=0}^{\infty} F_{yn} \cos(n\theta) + \sum_{n=0}^{\infty} F'_{yn} \sin(n\theta) \qquad [11.2]$$

$$F_z = \sum_{n=0}^{\infty} F_{zn} \cos(n\theta) + \sum_{n=0}^{\infty} F'_{zn} \sin(n\theta)$$

For the x and y axes, the cosine terms are symmetric, while the sine terms are antisymmetric about the $\theta = 0$ plane. However, the opposite is true for the F_z (circumferential) terms. The first two harmonic loads (*i.e.*

$n = 0$ and $n = 1$) for the three components are shown in Fig. 11.4. Note the symmetries and antisymmetries of each case. Higher harmonics ($n > 1$) will produce more complex variations around the circumference, but the same principles apply. In theory a model will require an infinite series to describe the loading exactly, but in practice it is only necessary to consider a limited number. More than ten harmonics are rarely required, and frequently fewer than five will be accurate enough. Figure 11.5 shows the loading distribution that is applied if five, ten and fifteen terms are used to represent a point load. The loading is certainly applied over a larger area when fewer terms are used, but if the load is distant from the particular area of interest then the approximation is acceptable. Many finite element systems will decompose the loading into harmonics for the user, and may even sum and display the terms.

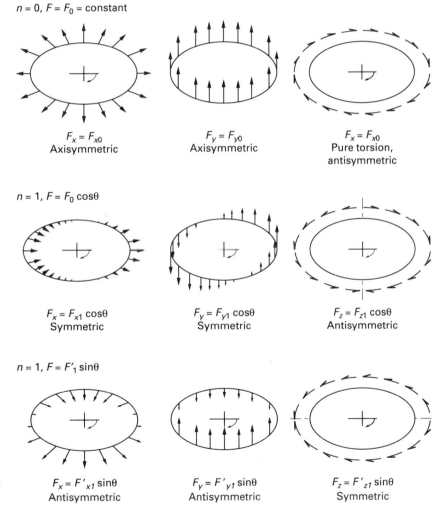

$n = 0$, $F = F_0 = $ constant

$F_x = F_{x0}$
Axisymmetric

$F_y = F_{y0}$
Axisymmetric

$F_x = F_{x0}$
Pure torsion,
antisymmetric

$n = 1$, $F = F_0 \cos\theta$

$F_x = F_{x1} \cos\theta$
Symmetric

$F_y = F_{y1} \cos\theta$
Symmetric

$F_z = F_{z1} \cos\theta$
Antisymmetric

$n = 1$, $F = F'_1 \sin\theta$

$F_x = F'_{x1} \sin\theta$
Antisymmetric

$F_y = F'_{y1} \sin\theta$
Antisymmetric

$F_z = F'_{z1} \sin\theta$
Symmetric

Figure 11.4 First two harmonic load components

The elements that are used for axisymmetric problems with non-axisymmetric loads are not the same as those described previously in Chapter 5. The special elements (sometimes referred to as harmonic elements) require a circumferential variation of the displacements, as well as the normal variation through the element. For example, the displacement in the radial direction might vary as

$$u = \sum_{n=0}^{\infty} u_n \cos(n\theta) + \sum_{n=0}^{\infty} u'_n \sin(n\theta) \qquad [11.3]$$

Therefore, since every harmonic component (*i.e.* every value of *n*) assumes a different distribution of the displacement through the elements, the stiffness matrix must be recalculated (and solved) for each load case. When compared with the single stiffness matrix calculation of a full three-dimensional analysis of the same problem, it might appear that harmonic elements are inefficient. However, the two-dimensional axisymmetric model will be considerably smaller than the three-dimensional version, and consequently a harmonic analysis invariably proves to be very efficient compared with other methods.

Once the individual responses of the harmonic loads are calculated, the post-processing of most commercial packages will automatically combine them and predict the displacements and stresses at any circumferential position.

Other types of symmetry and classes of problem

The application and processing of non-symmetric loads in symmetric problems can be quite awkward, but fortunately planar and axial symmetry modelling is supported in many systems. The same is not true, however, for cyclic and repetitive symmetry (Fig. 3.3) which are rarely supported, and require a considerable understanding of the method and behaviour of the problem to be solved correctly.

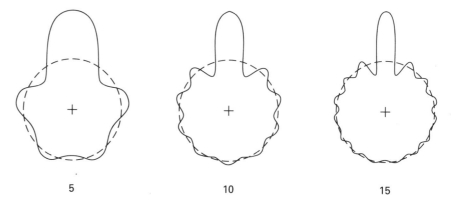

Figure 11.5 Applied loading when five, ten and fifteen harmonic terms are used to represent a point load

5   10   15

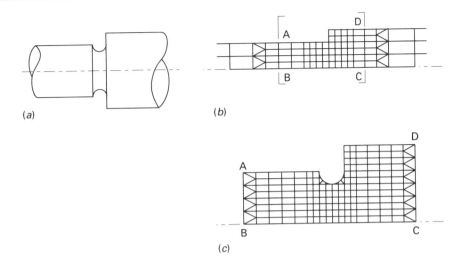

Figure 11.6 (a) Grooved
component (b) model of
component with coarse mesh
(c) model of groove with fine
mesh

Although the preceding discussion has concentrated on the application
of non-axisymmetric loading of axisymmetric stress problems, the ideas
can be applied quite easily to thermal analyses. As usual the thermal
calculations prove to be simpler than those performed in structural
analyses because of the scalar variable.

11.3   Submodelling and substructuring

Submodelling and substructuring are very useful methods for improving
modelling efficiency. Submodelling is used when a small detail needs to
be analysed in a large structure, and proceeds in two stages. Firstly, the
whole structure is modelled with a coarse mesh, and then the results of
this analysis are applied to a second model of the small detail using a fine
mesh. Consider the grooved axisymmetric component in Fig. 11.6. The
submodel contains the detailed geometry of the groove, where the nodes
around the edge of the fine model (*i.e.* along the lines AB and CD)
coincide with the nodes in the coarse model. The first stage of the analysis
determines the displacements of the boundary nodes in the complete
model, which are then applied to the groove model, and an accurate
stress analysis of the groove is thus obtained. However, note that for the
method to work the boundary nodes must lie outside the region of
influence of the groove.

Submodelling is advantageous because it minimizes the number of
degrees of freedom in a model. Hence, if computer resources are limited,
separating the fine mesh required to analyse a small detail from the main
model significantly reduces the model size. Also the risk of ill-
conditioning is minimized if the range of element sizes is limited in a

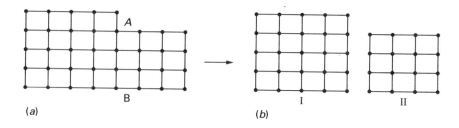

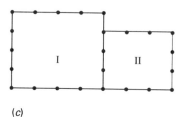

Figure 11.7 (a) Original
structure, divided into (b) two
substructures, thus creating
(c) two superelements

model. Finally, submodelling is particularly suitable if the component
detail needs to be examined many times, as is necessary in a parametric
study. For example, the groove in Fig. 11.6 could be optimized by simply
reanalysing the fine model and not the complete problem.

Substructuring is the analysis of a large or complex structure by
breaking it down into a number of smaller substructures. Consider the
example in Fig. 11.7, where the original structure of Fig. 11.7(a) is
divided into two substructures in Fig. 11.7(b). The substructures are
linked by the displacements (or temperatures) of the nodes on their
common interface (line AB), and the behaviour of the two parts can in
fact be expressed solely in terms of the behaviour of the nodes defining
their boundaries, as illustrated in Fig. 11.7(c). For example, if the
equations of the substructure are rearranged so that the displacements of
the boundary nodes are stored in vector $\{U_b\}$ and the displacements of
the remaining internal nodes are stored in vector $\{U_i\}$, they become

$$\begin{bmatrix} [k_{11}][k_{12}] \\ [k_{21}][k_{22}] \end{bmatrix} \begin{Bmatrix} \{U_b\} \\ \{U_i\} \end{Bmatrix} = \begin{Bmatrix} \{F_b\} \\ \{F_i\} \end{Bmatrix} \qquad [11.4]$$

The second row of Eq. 11.4 gives

$$\{U_i\} = [k_{22}]^{-1}\{\{F_i\} - [k_{21}]\{U_b\}\} \qquad [11.5]$$

and if this is substituted into the first row

$$[k_{11}]\{U_b\} + [k_{12}]\{U_i\} = \{F_b\} \qquad [11.6]$$

then simple rearrangement of the terms shows that the equations of the substructure can be expressed in the form

$$[k^E]\{U_b\} = \{F^E\} \tag{11.7}$$

Equation 11.7 defines the behaviour of the region in terms of the nodal displacements around its boundary, in the same way that individual finite elements do. The stiffness matrix has the features of all other stiffness matrices, for example it is symmetric. In fact the region can be considered to be a superelement; hence the superscript E used in Eq. 11.7.

If the two superelements in Fig. 11.7(c) are used to analyse the original structure, the results of the analysis in the first instance would be displacements of the boundary nodes. From these values, the displacements of the internal nodes and ultimately the stresses in the body could be derived from Eq. 11.5. Notice that the internal displacements and stresses need not be calculated in all the substructures, only those required. However, this does not necessarily save computer time, because the time consuming matrix inversion is required in any case to calculate the stiffness matrix of the superelement (*i.e.* in Eq. 11.5). Nevertheless, the technique does save considerable time if part of the original structure

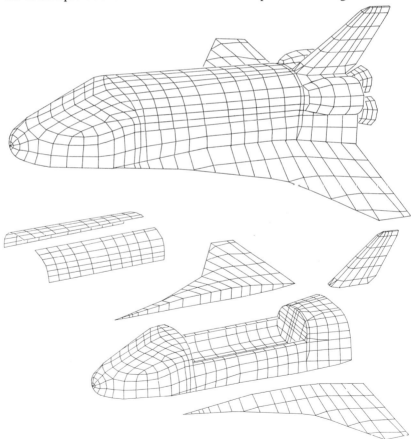

Figure 11.8 Finite element analysis of a space shuttle using substructuring

needs remodelling, but the design of other parts is finalized. For example, the superelement part I in Fig. 11.7(c) could be used time and time again for different designs of part II. This can result in significant savings when large models are analysed.

Another advantage of substructuring is that large problems can be analysed by several groups of engineers, with each group concentrating on one part of the structure. The only requirement is that the same nodal arrangement must be used on common interfaces. By the use of substructuring it is possible to generate finite element models of large complex problems, such as cars and aircraft. Figure 11.8 shows the type of finite element models used in the analysis of a space shuttle.

11.4   Element and program validation

It is emphasized many times in earlier chapters that the finite element method is approximate, and that its accuracy depends not only on the mathematical assumptions inherent in the method, such as the choice of interpolation function, but also on the reliable implementation of the algorithms in the computer software, such as the correct treatment of the prescribed loading and displacements. Furthermore, the size and complexity of the software packages mean that the essential mathematical routines are quite remote from the user and largely beyond his or her control. For these reasons it is vital that the user, and indeed the engineering community in general, are able to confirm the accuracy and reliability of the software. To prove the programs and gain confidence in the results, the behaviour of the elements needs to be examined individually and in an assembly. The following are just some of the ways in which this can be achieved:

(a)   single-element tests
(b)   patch tests
(c)   benchmark tests.

When the software developers and users have agreed on a series of assessment tests, it is possible that the program documentation will contain the results of such tests, and the models included with the software, to allow the user to validate the implementation on his own particular machine. (It is important to emphasize again, however, that the results would indicate only the quality of the implementation, and that it is the responsibility of the user to apply the software correctly.) The development of such a series of tests is one of the aims of NAFEMS (the National Agency for Finite Element Methods and Standards), and the notes below include some of these ideas.

Single-element tests

As the name suggests, these tests involve the examination of a single element. For example, the test might involve the analysis of a two-dimensional solid element under a range of loading and boundary conditions, as illustrated in Fig. 11.9. The results are compared with either classical solutions or standard finite element solutions using a large number of elements. For an element to be considered satisfactory, it must be capable of predicting a constant stress field, including a constant zero-stress field (which represents rigid body motion). However, while it might be easy to develop these conditions for an isolated rectangular element as illustrated in Fig. 11.9, the equivalent analysis of triangular elements and general quadrilaterals is not so easy.

Another approach is to consider the element as part of a larger continuum which experiences the constant stress condition, as illustrated

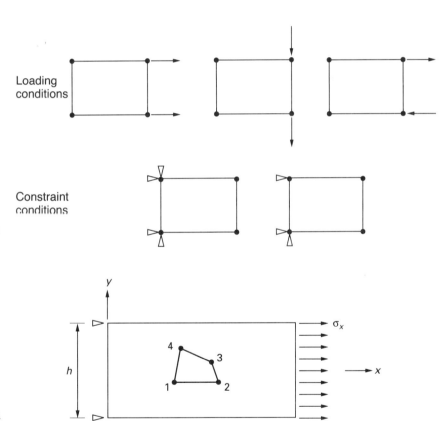

Figure 11.9 Possible loading and constraint conditions for a simple single-element text

Figure 11.10 Constant stress loading of a rectangular region containing the element under investigation

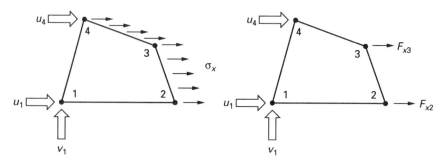

Figure 11.11 Constant σ_x stress conditions for the single element

in Fig. 11.10. The displacement at any part due to the applied stress can be calculated from

$$u = \frac{\sigma_x}{E}x$$

$$v = -v\frac{\sigma_x}{E}(\frac{h}{2} + y)$$

[11.8]

The element is then considered in isolation, and the correct combination of enforced displacements and nodal loads is applied to it in order to develop the constant stress field. For example, if the enforced displacements are applied to nodes 1 and 4, the element would be loaded as shown in Fig. 11.11(a). The stress (pressure) loads on the element faces are converted to nodal forces in the usual way to give the final test conditions, as in Fig. 11.11(b).

The analysis of other stress fields (*i.e.* σ_x and τ_{xy}) proceeds in the same way, and the investigation of different element types can be achieved by a similar approach. Clearly this type of single-element test is ideal for examining the effects of element distortion.

Patch tests

A patch test is the analysis of an assembly of elements subjected to a constant stress field, and the elements successfully pass the test if they all contain the same constant value of applied stress. The type of patch that might be used is shown in Fig. 11.12. The left-hand edge is constrained, and nodal forces or enforced displacements (calculated from Eq. 11.8) are used to load the model. By sensible mesh design the patch test and the (second) single-element test can be conveniently linked, since the single element can be part of the patch.

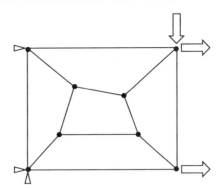

Figure 11.12 Simple patch test model

Benchmark tests

The fundamental element tests outlined previously are suitable for confirming basic element behaviour and examining the effects of element distortions. However, the tests are limited, and indeed some elements (for example those using reduced integration schemes) may fail such tests and yet perform quite satisfactorily in 'real' problems where the number of elements is that much larger. Moreover, single-element and patch tests only verify limited features of the software. Benchmark tests are designed to analyse more realistic problems, and to investigate not only the accuracy of implementation of the routines, but also the quality and general robustness of the complete program, including the pre- and post-processors. For example, apart from the obvious requirement that the pre-processor should generate an accurate mesh, it is desirable if it warns the user of badly distorted elements or inadequately restrained models. Similarly, the post-processor must calculate and present the results correctly, using accurate interpolation routines. As pre and post-processors have become more complex with an ever increasing number of sophisticated facilities, the user has lost contact with the actual finite element routines. Although the sophistication provides invaluable assistance to the user, it does mean that many programs can be used as 'black boxes', easily producing results that appear professional and visually convincing. Because of this effect, all programs should perform a minimum number of basic checks on the model.

Two tests proposed by NAFEMS are shown in Figs 11.13 and 11.14. The first of these is a linear elastic problem, and involves the pressure loading on an elliptic membrane; the second is a one-dimensional transient heat conduction problem. Both the tests are devised to test program accuracy. They are clearly specified, and require a single target value.

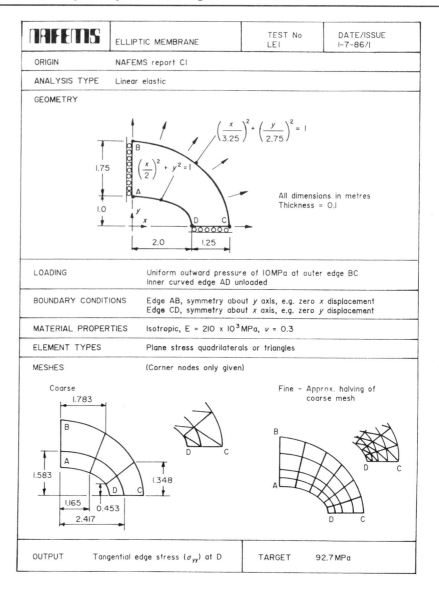

Figure 11.13 Linear elastic benchmark test (courtesy of NAFEMS)

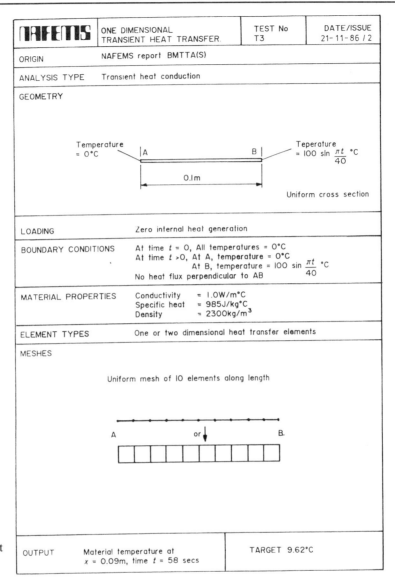

Figure 11.14 Transient heat conduction benchmark test (courtesy of NAFEMS)

12 Commercial finite element programs

12.1 Introduction

There are numerous commercial finite element packages of varying complexity available at present. Many programs will have taken hundreds of man-years to develop and will have teams of engineers and programmers constantly improving and updating them. Thus there is little point in an individual engineer writing his own system, unless it is for a very specific task. In this chapter the main features that can be expected in commercial programs are described, but the discussion concentrates on static and steady state modelling and is by no means exhaustive. As will become obvious, much effort is directed towards the development of the pre- and post-processors for the programs, and undoubtedly the selection of many finite element systems is based on their facilities rather than the essential finite element routines.

The example commands used in this chapter are for illustrative purposes only, and are not taken from any particular finite element program.

12.2 General facilities

Most programs are now available in modules, allowing the prospective user to select only those parts of the software he requires. However, in practice most of the modules use the same pre- and post-processors, and it is only the solution phases that are different.

Documentation is a vital part of any computer program, and is especially important for finite element software. A well supported finite element package will be supplied with several basic manuals, such as a user manual, a theoretical manual and a verification and sample problem manual. However, since the user manual can be very large and detailed, introductory guides are particularly useful for new users. There is an increasing move to include on-line help facilities with the software, and, as media storage improves, it can be expected that this on-line help will become more detailed and make some of the paper manuals redundant. The verification and problem manuals are also important for new users, since they give examples of the commands necessary to analyse sample problems. Large problems can have hundreds of commands, and without such examples it can be very time consuming for a user to set up a new problem type.

The range of computers supported by the finite element software covers most available types, from basic personal computers (PCs) through minis and mainframes to supercomputers. PCs are becoming increasingly powerful and readily available, and are widely supported by most general packages. In particular 386-based PCs, usually with extended random access memory (RAM), can solve realistic problems with limitations on the model size dictated only by the available disk space. The more powerful machines are necessary for large and/or complex problems, such as transient and non-linear analyses.

Dedicated computers such as PCs and workstations are usually operated interactively. Larger general-use machines may require at least the solution phase to be executed in batch mode, possibly overnight, since the finite element routines can be very intensive in both processing terms and disk space requirements. The pre- and post-processors of an increasing number of packages are menu driven rather than command driven, which results in a more user friendly interface and a generally more robust system. The selections are made either through the keyboard, or by selecting with cross-hairs moved by a mouse, a graphics tablet or a joystick. The screen layout of such a menu-driven finite element program is shown in Plate 13.

When the finite element method was first applied, a model was defined as a long list of numbers, and the results appeared as an even longer list. Nowadays, however, the engineer generates the model and views the results graphically, and the plotting facilities of the programs are among their most important features. Generally models (and results) can be rotated and viewed from any angle, possibly with perspective and light source shading, and the user can zoom in and out to examine particular details. Hidden line plots are also usually available (see Plate 5), together with section plots which allow the performance of any part of the problem to be fully examined. (Specific pre- and post-processor facilities are discussed in the following sections.)

Having produced the plots on the screen, it is of course desirable to get a hard copy. The easiest way is to perform a screen dump on to a suitable printer, but the resolution will be limited to the resolution of the screen or the printer software. The best quality output is produced by devices which are specifically supported and driven by the software.

12.3   Pre-processors

The pre-processor is used to develop the finite element model. There are essentially two methods in which the element mesh can be generated: direct user input or automatic mesh generation. In the first of these the user defines individual nodes and elements and builds up the mesh manually, while in the second a solid model of the problem is developed and the computer then generates the mesh automatically. The latter method also allows CAD drawings to be used to define the solid model. Some features are obviously common to both approaches, and indeed most programs allow the two methods to be intermixed.

12.3.1 Direct user input

Coordinate systems

The ability to define different types and numbers of coordinate systems can simplify mesh generation and the application of the boundary conditions enormously. Cartesian, cylindrical and spherical systems are commonly available for both global and local systems (Fig. 12.1), allowing very flexible model generations. For example, the definition of the coordinates of a series of nodes on a circular arc is much easier in a cylindrical than in a Cartesian system (Fig. 12.2). Furthermore, if the origin of the different coordinate systems can be defined in any position in space (*i.e.* local systems are available), even apparently complex patterns can be produced. The pre-processor simply converts the coordinates in the local coordinate system to the global Cartesian one. As a rule, when a user defines a coordinate system it becomes the 'active' system, and subsequent commands are referenced to that system, until another one is declared or an old one is reselected. (The coordinate systems are usually numbered to aid in identification.)

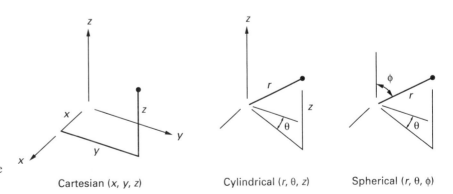

Figure 12.1 Various coordinate systems

Cartesian (*x*, *y*, *z*)   Cylindrical (*r*, θ, *z*)   Spherical (*r*, θ, φ)

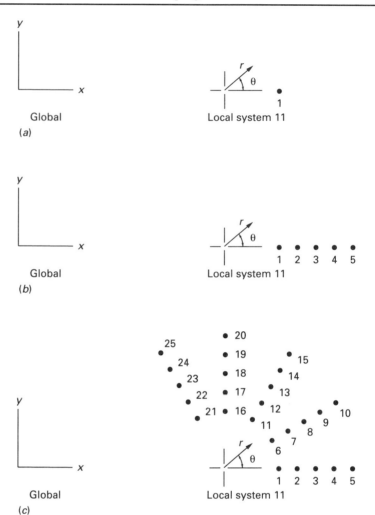

Figure 12.2 Typical commands to generate a pattern of nodes using a local cylindrical coordinate system (r, θ, z):

(a) NODE,1,5,0,2 (= define node 1 at $r = 5$, $\theta = 0$, $z = 2$)

(b) NODE-GEN,4,1,,,2,0,0 (= generate 4 nodes from node 1, incrementing r by 2, and θ and z by 0)

(c) NODE-GEN,4,1,5,1,0,30,0 (= generate 4 patterns of nodes from nodes 1 to 5 (step 1), incrementing r by 0, θ by 30 and z by 0)

Nodes

Once one node has been defined it can easily be developed into a line of nodes by incrementing one or more of its coordinates; similarly, a line of nodes can be generated into an area and ultimately a volume. Figure 12.2 shows the generation of such a nodal pattern by the use of just three commands and a local cylindrical coordinate system (labelled number 11). Similar commands are also usually available that can translate and mirror node patterns about user defined axes. Clearly, the pre-processor is just performing a series of very simple but repetitive calculations for the user.

Elements

When the nodes of the model have been generated, the next step is to define the elements. The actual selection of the element type is discussed elsewhere in this book, and obviously needs to have been made before

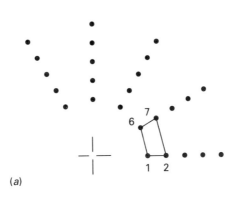

(a)

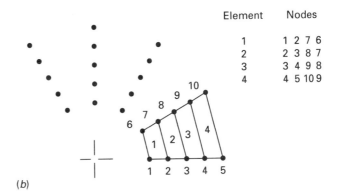

Element	Nodes
1	1 2 7 6
2	2 3 8 7
3	3 4 9 8
4	4 5 10 9

(b)

Figure 12.3 Typical commands to generate a pattern of two-dimensional four-noded elements:

(a) ELEMENT,1,2,7,6 (= define an element using nodes 1, 2, 7 and 6, in an anticlockwise direction; element number automatically chosen to be 1)

(b) ELEMENT-GEN,3,1,,,1 (= generate 3 elements from element number 1, by incrementing the node pattern by 1)

(c) ELEMENT-GEN,3,1,4,1,5 (= generate 3 patterns of elements from elements 1 to 4 (step 1) by incrementing the node pattern by 5)

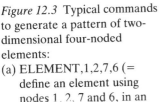

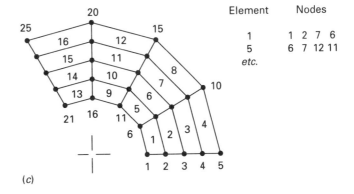

Element	Nodes
1	1 2 7 6
5	6 7 12 11
etc.	

(c)

definition of the nodes can begin. Patterns of elements can usually be generated in a similar way to that used for the nodes. Consider for example the definition of the twenty (two-dimensional) elements in Fig. 12.3; again just three simple commands are required. For such two-dimensional elements it is vital that the nodes are specified in a consistent manner, either clockwise or anticlockwise. The generation commands for defining both nodes and elements are very powerful, and can build two- and three-dimensional meshes very quickly. The simple example shows, however, that it is important to plan the node patterns carefully and in a logical manner, so that the elements can be specified with the minimum of effort.

Geometric constants

Some elements require more information than can be gained solely from the nodal geometry. For example, the user must specify the thickness of plane stress elements, and the area and second moment of area of beam elements.

Material constants

Materials are assumed to be isotropic unless the user specifies otherwise, and not to be a function of temperature unless the user inputs the variation of the properties with temperature. Where a non-constant value is to be used, the data can be input in tabular form or possibly by specifying a polynomial variation. Some programs will also plot the data once specified by the user.

Loads

When the nodes and elements have been generated, the loads must be applied to the model. For stress problems this involves specification of nodal forces, pressures, body forces or enforced displacements. Nodal forces are defined in the active coordinate system, and resolved into the global system by the program. Thus the force of 100 N applied to nodes 22 to 24 in Fig. 12.4(a) is most easily specified in the local Cartesian coordinate system (labelled number 12).

The application of pressure loads can usually be achieved by a variety of commands, and the final choice will depend on the easiest way of specifying the pressure surface. For example, the simplest method is to define the side of the element by its node numbers, as in the first command in Fig. 12.4(b). However, if the pressure acts on many element faces which lie at a constant distance from the (active) origin, then

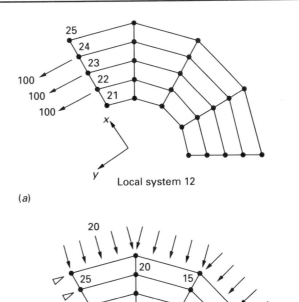

(a)

Figure 12.4 Typical commands for the application of (a) nodal loads, using local Cartesian coordinate system 12 (b) pressure loads, using local cylindrical coordinate system 11. Examples:
(a) FORCE,22,y,100 (= apply a force at node 22 in the *y* direction of size 100)
(b) PRESSURE,5,10,r,−20 (= apply a pressure between nodes 5 and 10 in the *r* direction of −20);
PRESSURE-SURF,r,13,−20 (= apply a pressure on all element faces along *r* = 13 of −20)

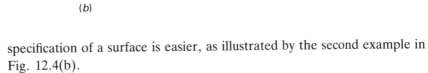

(b)

specification of a surface is easier, as illustrated by the second example in Fig. 12.4(b).

Care must be taken when specifying gravitational or other accelerations to ensure that the units are compatible. If the dimensions, acceleration and material density are in standard SI units (*i.e.* metres, kilograms and seconds) then body forces are calculated correctly, but models defined in dimensions other than metres are frequently prone to errors due to the incorrect specification of the density. Remember that

$$\text{force} = \text{mass} \times \text{acceleration} = \text{density} \times \text{volume} \times \text{acceleration}$$
$$= \text{kg} \times \text{m/s}^2 = \text{Mg} \times \text{mm/s}^2$$

Thus when a model is generated in millimetres, the densities of the materials need to be specified in units of Mg/mm^3.

Loading in thermal problems involves convection, incident heat flux and heat sources. As with applied pressures, the surface associated with any convection or heat flux can usually be specified by either node number or coordinate position. The heat sources are defined at individual nodes in the same way as nodal forces, except of course that there is no need to define a direction.

If the geometry and material conditions remain constant, then a series of stress and thermal load cases can be analysed for a relatively small extra cost, because the stiffness matrix remains the same. Consequently, the option of specifying various load runs is available in most pre-processors, and the corresponding facility of combining different load cases is available in post-processors.

Constraint conditions

Local coordinate systems are particularly helpful in the definition of constraint conditions in stress problems, where the relevant element faces are inclined to the global coordinate system. Nodal constraints can be applied in the same way as the nodal forces. For example, if the model in Fig. 12.4(b) was constrained in the hoop direction at nodes 21 to 25 then a command of the form

```
DISPLACEMENT,21,theta,0
```

would set the displacement in the theta direction at node 21 equal to zero (in the local cylindrical coordinate system 11). The same command could also be used to apply an enforced displacement. The restraint of nodes 21 to 25 as just described could indicate that they lie on a line of symmetry, and thus there is no movement perpendicular to that line. Some programs allow the user to directly specify the line of symmetry and then calculate the necessary constraints automatically. A few will also calculate the antisymmetry constraints as well, if necessary.

For thermal analyses, such a symmetry condition implies that no heat flows across the line of symmetry. Thus the edge of the model along the symmetry line is adiabatic. This does not require the application of any boundary condition, because if there is no connecting model there will be no heat flow. The specification of an individual nodal temperature in a model is straightforward. For example, the following command would result in the temperature at node 1 being set at 100:

```
TEMPERATURE,1,100
```

Since temperature is a scalar, no direction is needed.

List, delete and plot facilities

Vital functions of any pre-processor program are the abilities to list, delete and plot the finite element model details. The importance of checking the model is discussed in detail in Section 9.5, and some of the features illustrated include hidden line plots and shrunken element plots. Most programs will plot the nodes, elements, loading and constraint conditions applied to the model, together with the position and orientation of any user defined coordinate systems. Because of the size and complexity of some finite element models, it is also usually possible to select a subset or range of most items prior to listing or plotting, allowing the data to be checked that much more easily.

Deleting any part of the model is an important facility. Its purpose is not only to allow the user to correct mistakes. In some cases it is easier to generate a uniform mesh of elements and then delete a region, rather than to try to mesh around an awkward feature such as a hole.

The analysis file

The last phase of pre-processing is the preparation of the analysis file for input into the finite element solution program. Pre-processors normally check the data at this point to identify badly distorted elements and potential modelling errors. Element distortion is discussed in detail in Section 9.4, including the type of checks that the program can perform. Most programs issue warnings for elements that are badly distorted, and some stop with an error if severe distortion is detected.

Other useful facilities that are available include warnings of unused nodes and multiple definitions of elements, *i.e.* where the user has defined an element twice. The program cannot of course check if an element has been omitted, but a detailed examination of the element plots with shrink and hidden line options usually results in identification of gaps in the model (Fig. 9.33 and Plate 5).

Before the analysis file is written, some programs require the user to specify any special output data he needs for the post-processing phase, such as the reactions or the strain energy density. Otherwise the calculations are kept to a minimum, and only the basic output quantities (displacements and stresses or temperatures and heat flows) are produced.

12.3.2 Automatic mesh generation

Instead of requiring the nodes and elements to be defined directly, many programs offer automatic mesh generation facilities. The user defines the region to be meshed and selects the type of element to be used, and the computer then devises the mesh. Apart from specifying the layout and number of elements along the edges of the region, the user has little control of the process. Automatic mesh generation requires a solid modeller with which the user builds up the geometry of the problem. The shape is constructed by defining points, lines, areas and volumes, and again local coordinate systems prove to be invaluable. For example, consider the automatic mesh generation of the problem in Fig. 12.5. The user defines points at the corners of the region and lines connecting the points, as in Fig. 12.5(a). (Note that since cylindrical coordinate system 11 is the active system, lines L2 and L4 are generated as arcs and not straight lines.) The area is then specified by the area command. Some mesh generators allow the area to be defined directly from the corner

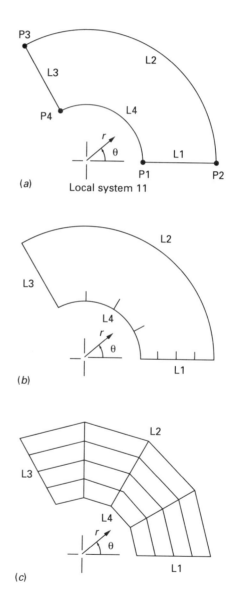

Figure 12.5 Typical commands used for automatic mesh generation: (a) geometry definition (b) specification of mesh layout (c) automatic mesh of area 1, using four-noded quadrilateral elements. Examples: (a) POINT,1,5,0,2 (= define point 1 at $r = 5$, $\theta = 0$, $z = 2$); LINE,3,3,4 (= define line L3 between points P3 and P4); AREA,1,1,2,3,4 (= define area A1 by lines L1, L2, L3, L4) (b) LINE-DIV,1,4 (= divide line 1 into 4 element sides); LINE-DIV,4,8 (= divide line 4 into 8 element sides) (c) AREA-ME4H,1 (= mesh area 1)

points, and automatically fill in the lines (L1 to L4). When the region has been specified, details of the element type and layout are input as in Fig. 12.5(b) and the area is then meshed by one command as shown in Fig. 12.5(c).

It is usually possible to repeat points, lines, areas and volumes in the same way that node and element patterns are repeated earlier in the chapter. Furthermore, some programs will automatically calculate the intersection of curved surfaces, as for example in the intersection of two circular pipes, and also generate fillet radii automatically for the user.

When the mesh has been generated, the loading and constraint conditions need to be applied as usual. This can be achieved either by reference to individual nodes and element faces as described previously, or by using the points, lines and areas again. (For example, in Fig. 12.5 a pressure could be defined on line L2, or a force on point P1.)

Parametric input is available in many systems, allowing the user to define the model in terms of a limited number of parameters, which can then be easily changed to develop a different version of the model. Consider the model in Fig. 12.5(a). If the coordinates of the points are defined parametrically, different sized models can be analysed very quickly. For example, if two parameters *A and *B are used so that

```
*A = 5
*B = *A + 8
point,1,*A,0,2
point,2,*B,0,2
```

then changing *A is all that is required to develop a different sized model.

There is little to be gained by using automatic mesh generation for the problem in Fig. 12.5 because the layout is so simple, but for more complex shapes it can prove to be a very useful facility. It is generally easier for mesh generators to use triangular and tetrahedral elements than quadrilateral and hexahedral elements, and consequently some meshing programs only allow the use of rectangular faced elements on simple geometries. However, as the sophistication of the meshing algorithms improves, this limitation will no doubt be overcome. An excellent example of the capabilities of automatic mesh generators is shown in Plate 14.

12.4 Post-processors

After the finite element equations have been assembled and solved, the user then begins the post-processing phase. For very large models or smaller models with many iterations, the amount of output data can be vast, and is frequently too large to be handled all at once. The simplest form of post-processor merely reads in the analysis results and presents them to the user. This type of program actually has few processing facilities, and the output data requirements must be specified in the analysis file and are calculated directly after the solution phase. More sophisticated finite element software records all essential output information in a file and then works on the data through the post-processor. Like the better pre-processors, the range of facilities can be large. For stress problems, the essential output data are the nodal displacements and stresses at the Gauss integration points, and from these values the post-processor can extrapolate the stresses to predict the nodal values, and calculate the principal and Von Mises stresses.

Most output data can usually be printed and plotted with the plotting facilities previously described for the pre-processor (*i.e.* model rotation, zooming and hidden line plots). A very useful option available with some programs is the ability to define local coordinate systems in the post-processor, and thereby to output the results in any orientation. Also, routines which sort the output data and list the results in descending order can save the user a significant amount of time.

Probably the most useful plots are the deformation and stress contour plots (Plate 2), but principal stress and heat flow vector plots can also give considerable insight into the load paths (Plate 4). For three-dimensional problems, the ability to produce plots at any section through the model can also prove to be invaluable.

Finally, many programs now offer the option of animation. By scaling the displacements from −100 per cent (or zero) to 100 per cent of the predicted values, and plotting the values sequentially on the screen, an animated view of the loading can be created, showing how the structure deflects.

Bibliography

Textbooks

AKIN J E 1982. *Application and Implementation of Finite Element Methods* Academic Press

AKIN J E 1987. *Finite Element Analysis for Undergraduates* Academic Press

BARKER A J, PEPPER D W 1991. *Finite Elements 1-2-3* McGraw-Hill

BREBBIA C A 1982. *Finite Element Systems: A Handbook* Springer-Verlag

BROWN D K 1984. *An Introduction to the Finite Element Method Using Basic Programs* Surrey University Press

BURNETT D S 1987. *Finite Element Analysis: From Concepts to Applications* Addison-Wesley

CHEUNG Y K, YEO M F 1979. *A Practical Introduction to Finite Element Analysis* Pitman

CHUNG T J 1978. *Finite Element Analysis in Fluid Dynamics* McGraw-Hill

CONNOR J J, BREBBIA C A 1976. *Finite Element Techniques for Fluid Flow* Newnes-Butterworths

COOK R D 1974. *Concepts and Applications of Finite Element Analysis* J Wiley

CRISFIELD M A 1991. *Non-linear Finite Element Analysis of Solids and Structures* J Wiley

DAWE D J 1984. *Matrix and Finite Element Displacement Analysis of Structures* Oxford University Press

DESAI C S, ABEL J F 1972. *Introduction to the Finite Element Method* Van Nostrand Reinhold

DHATT G, TOUZOT G, CANTIN G 1984. *The Finite Element Method Displayed* J Wiley

FENNER R T 1975. *Finite Element Methods for Engineers* Macmillan

GOULD P L 1985. *Finite Element Analysis of Shells of Revolution* Pitman

GRANDIN H 1986. *Fundamentals of the Finite Element Method* Macmillian

HINTON E, OWEN D R J 1977. *Finite Element Programming* Academic Press

HUEBNER K H, THORNTON E A 1982. *The Finite Element Method for Engineers* J Wiley

HUGHES T J R 1987. *The Finite Element Method: Linear Static and Dynamic Finite Element Analysis* Prentice-Hall

IRONS B, SHRIVE N 1983. *Finite Element Primer* Ellis Horwood

KIKUCHI N 1986. *Finite Element Methods in Mechanics* Cambridge University Press

LIVESLEY R K 1983. *Finite Elements – An Introduction for Engineers* Cambridge University Press

NORRIE D H, De VRIES G 1973. *The Finite Element Method: Fundamentals and Applications* Academic Press

OWEN D R J, HINTON E 1980. *A Simple Guide to Finite Elements* Pineridge Press

PAO Y C 1986. *A First Course in Finite Element Analysis* Allyn & Bacon

RAO S S 1989. *The Finite Element Method in Engineering* Pergammon Press

REDDY J N 1984. *An Introduction to the Finite Element Method* McGraw-Hill

RICHARDS T H 1977. *Energy Methods in Stress Analysis, With an Introduction to Finite Element Techniques* Ellis Harwood

ROSS C T F 1982. *Computational Methods in Structural and Continuum Mechanics* Ellis Horwood

ROSS C T F 1985. *Finite Element Methods in Structural Mechanics* Ellis-Horwood

SEGERLIND L J 1984. *Applied Finite Element Analysis* J Wiley

SMITH I M 1982. *Programming the Finite Element Method, with Application to Geomechanics* J Wiley

STASA F L 1985. *Applied Finite Element Analysis for Engineers* Holt-Sanders

WAIT R, MITCHELL A R 1985. *Finite Element Analysis and Applications* J Wiley

ZIENKIEWICZ O C, MORGAN K 1983. *Finite Elements and Approximation* J Wiley

ZIENKIEWICZ O C, TAYLOR R L 1989. *The Finite Element Method* McGraw-Hill

ZIENKIEWICZ O C, TAYLOR R L 1991. *The Finite Element Method: Volume 2 – Solid and Fluid Mechanics, Dynamics and Non-linearity* McGraw-Hill

A finite element primer. National Agency for Finite Element Methods and Standards, Dept of Trade and Industry, 1986

Relevant journals and periodicals

Benchmark The magazine of NAFEMS (National Agency for Finite Element Methods and Standards), NAFEMS, Glasgow

Computers and Structures Pergammon Press

Finite Element News Robinson and Associates, Okehampton, Devon

Finite Elements in Analysis and Design The International Journal of Applied Finite Elements and Computer Aided Engineering, North-Holland

Glossary

This glossary refers in particular to finite element analysis, but note that some terms may have other meanings in other contexts.

Accuracy: The closeness of the solution predicted by a finite element model to the correct answer of the problem

Anisotropic material: A material with different mechanical properties in three mutually perpendicular directions

Antisymmetric loading: Where the load profiles and points of application (on a symmetric body) are symmetric, but the load directions are not symmetric

Approximating function – see interpolation function

Area coordinate: A natural coordinate for a triangular element with a value between zero and unity, equal to the ratio of a triangular portion of the element to its total area

Aspect ratio: A measure of the distortion of an element, derived from the ratio of two representative dimensions of the element

Assembly of the element equations: The addition of the individual element equations to form the system equations of the problem

Automatic mesh generation: A process where the computer generates the node and element details for a finite element model from a 'solid model' of the problem

Axial symmetry: A symmetry condition where the properties of the radial sections of a body of revolution are identical

Axisymmetric element: A two-dimensional element formulated for use in problems with axial symmetry. When the elements are used only a radial section of the geometry needs to be modelled

Banded matrix: A matrix where all non-zero terms are found to be confined to a band about the main diagonal of the matrix

Bandwidth: The width (or half-width) of the band of a banded matrix

Beam element: An element formulated to model beam type structures, with translational and rotational degrees of freedom at each node. Thick and thin beam formulations may be available

Benchmark test: A standard test model devised to examine the performance of finite element programs

Body force: A force produced by an acceleration effect, such as that due to inertia and gravity

Boundary conditions: The loading, constraints and other external effects applied to a model

Boundary line plot: A computer generated plot of the boundaries or edges of a model

Brick element: A solid, three-dimensional element with a tetrahedral, wedge or hexahedral shape

Buckling: Where a structure suddenly deforms out of plane at a critical load

Calculus of variations: A branch of mathematics concerned with finding the stationary values (e.g. minima and maxima) of an integral with respect to a function

Cholesky decomposition: A method of solution of the system equations where the stiffness matrix is decomposed into upper and lower triangular matrices

Compatible element – see conforming element

Completeness: A convergence condition where the field variable and a number of its derivatives must assume a constant value in an element as the size of the element decreases

Complex element: An element with a quadratic, cubic or higher interpolation function, having the same shape as the equivalent simplex element

Composite material: A material made of a number of discrete constituents, and in particular, laminated forms where different materials, such as glass and carbon fibre reinforced plastics, are layered with varying orientations

Conforming element: An element that satisfies both the completeness and continuity convergence conditions

Connectivity: Nodal information for the individual elements, and details of how they fit together to form the complete model

Consistent mass matrix: One type of mass matrix for a structural element, where the mass is distributed accurately through the element (cf lumped mass matrix)

Constant strain/stress element: An element for stress analysis problems formulated with a linear variation of displacement, resulting in a constant value of strain and stress across the element

Constraint: Any restriction imposed on the value of the (unknown) field variable. For example, a specified displacement for stress problems or temperature for thermal problems

Constraint equation: A relationship linking the behaviour of the degrees of freedom of two or more nodes

Continuity: A convergence condition where the field variable and a number of its derivatives must be continuous across element boundaries

Contour plot: The presentation of the state of stress (for example) in a model by a series of lines of constant stress, or different coloured bands defining different ranges of stress

Convergence: The improvement in the accuracy of an analysis by an increase in the sensitivity (ie accuracy) of the finite element mesh (by h- or p-refinement)

Coordinate transformation: The transformation of a matrix or vector from one coordinate system to another

Creep: The time dependent deformation of a component, even though the induced stresses are below the yield stress of the material

Curvilinear coordinate system: A local non-orthogonal coordinate system, used in particular to define multiplex elements

Cyclic symmetry: A symmetry condition where the same pattern is seen to be repeated sequentially about an axis

Degree of freedom: A variable used to describe the behaviour of a node in a body. For example, displacement or rotation in a stress problem, and temperature in a thermal problem

Direct integration solution method: A (time domain) method of solution for transient problems involving a timewise step-by-step approximation of the equations of motion

Direction cosine: Used in the specification of a direction. It is the cosine of the angle between the direction and a coordinate axis

Discretization: The division of the geometry of a problem into nodes and elements

Displacement pass: The calculation of the displacements of the master degrees of freedom in a dynamics analysis

Displacement plot: A computer generated plot of the distorted shape of a loaded component. Displacement plots are usually scaled for display, and often overlayed on the original geometry

Displacement response spectra: Graphs showing the displacement of a single degree of freedom system (with a range of natural frequencies and damping ratios) to a specified shock load

Distortion (of an element): Where an element is defined with a non-ideal shape. For example, the best shapes for quadrilateral and triangular elements are square and equilateral respectively

Duplicate element: This occurs if one element is accidentally defined on top of another element

Dynamic problem: A structural mechanics problem where the forces and displacements are a function of time

Edge plot – see boundary line plot

Eigenvalue problem: Where a critical value (or values) of a variable results in the non-trivial solution of a problem. For example, a critical buckling load in a buckling analysis, or a natural frequency in a modal analysis

Eigenvector: The vector of unknown degrees of freedom in a problem corresponding to a calculated eigenvalue. For example, the displacements in buckling and modal problems, but note that the magnitudes of the displacements are normalized

Element: The basic sub-division used to represent a body, with an associated interpolation function which approximates the distribution of the unknown field variable through the element

Element type: A general categorization of elements by their characteristics, for example, behaviour, shape, or material properties, into classes such stress, thermal, beams, solid and composite

Element validation: The verification of the performance of an element by the use of simple, single element and patch tests

Enforced displacement: The specification of a particular value for a nodal displacement. Most often a value of zero is specified, ie the node is constrained

Field interpolation function – see interpolation function

Field problems: A class of problems governed by the same basic differential equation, namely the field equation. Examples of field problems are heat transfer, irrotational flow, and Prandtl's torsion method

Finite strain: When the strain experienced by an element is considered to be so large that the original stiffness matrix is no longer representative

Follower loads: Loads which maintain the same relative orientation to a body as it undergoes large deformation

Forcing function: A time varying load applied to transient problems

Free vibration: The vibration of a structure or component without any external loading

Frequency domain solution: A method of solution for transient dynamic problems where the forcing function is decomposed into its frequency components and the solution then found by summing the effects of these components

Frequency sweep: A series of harmonic response analyses over a range of frequencies allowing the generation of graphs of frequency versus nodal displacement, velocity, acceleration or stress

Frontal solution method – see wavefront solution method

Functional: A function of several other functions. In particular, a functional is derived from the governing differential equation and boundary conditions of field problems, and used in the derivation of the field finite element equations

Galerkin's method: A weighted residual method for calculating the finite element equations of a problem where the weighting functions equal the shape functions

Gap element: A special element used to represent the contact between two parts of a model. The element can only transmit a compressive direct force, but might include a friction force

Gauss quadrature: A method of numerical integration where the value of the integrand is evaluated at sample points and then summed after factoring with weighting functions. If applied accurately with the correct number of sampling points, Gauss quadrature will give the exact answer

Gaussian elimination: The standard method of inverting a matrix (the stiffness matrix) to yield the unknown nodal values of the field variable

Geometric non-linearity: Where the geometry of the problem changes significantly so that a linear analysis is no longer acceptable. This might be due to large displacements or stress (geometric) stiffening

Geometric stiffening – see stress stiffening

Geometry approximation: An approximation of the full three-dimensional nature of a problem by a one-dimensional, two-dimensional or symmetry model

Geometry interpolation function: A function to approximate the geometry of an element in the same way that the field variable is approximated in the element. Elements with straight sides have linear geometry interpolation functions, those with curved sides have quadratic or higher order functions

Global coordinate system: The reference coordinate system for individual elements and the complete model, to which all other coordinate systems are referred

Gradient vector: A column vector containing the gradients of the field variable in field problems. Equivalent to the strain vector in stress problems

Guyan reduction: A method of condensing out the slave degrees of freedom from the system equations to leave the master degrees of freedom

h-refinement: An improvement in the accuracy of a finite element mesh by increasing the mesh density

Harmonic element: A special axisymmetric element used for the analysis of axisymmetric problems with non-axisymmetric loads. This element is not the same as the standard axisymmetric element used where the load is symmetric

Harmonic load (for harmonic response analyses): A load that varies sinusoidally in time at a single frequency

Harmonic load (for non-axisymmetric load cases): One of a series of loads used to approximate non-axisymmetric loading for harmonic elements, obtained by the Fourier method. The solution to the original problem is calculated by combining the responses of the individual loads

Harmonic response analysis: The analysis of a component or structure when subjected to a set of harmonic loads of known amplitude and frequency

Hidden line plot: A computer generated plot of a three-dimensional model where element details obscured by other elements are not shown, resulting in a proper three-dimensional view of the model

Higher order element: Any element with a second order or higher (ie non-linear) interpolation function

Hyperelastic material: A material that can withstand large, finite strains without exceeding the yield stress of the material, such as rubber

Ill-conditioning (of the stiffness matrix): This occurs when there is a large difference in the magnitude of the terms in the stiffness matrix, making the solution susceptible to round-off errors

Incompatible element – see non-conforming element

Incremental solution: Where the loading is applied in steps to the model, and the values predicted from each load step are then used to modify the element equations before the next load step is applied

Initial strain: Where an element is strained by effects other than the stresses developed due to external loading. For example, thermal strain which is caused by a change in temperature

Integration point: A sample point in an element at which the element equations are evaluated during numerical integration

Interface element – see gap element

Internal angle: The angle measured between the sides of a two- or three-dimensional element, and used as a guide to the distortion of the element

Internal node: A node inside an element, ie not a corner or mid-side node

Interpolation function: A function used to approximate the distribution of an unknown (field) variable through an element

Isoparametric element: An element where the geometry interpolation function and field interpolation function are of the same order

Isotropic material: A material with the same mechanical properties in three mutually perpendicular directions

Isotropic hardening rule: A method of describing how a material yields during plastic deformation

Iterative solution: A solution requiring several iterations to reach a solution that is consistent with the problem formulation. For example, iterative solutions are required with gap elements which can only transmit compressive loads, and with elastic–plastic problems to ensure that the stress–strain relationships are correct

Jacobian: This is frequently used as shorthand for the determinant of the Jacobian matrix

Jacobian matrix: A matrix containing the derivatives of the global coordinates with respect to the natural coordinates of a particular element

Kinematic hardening rule: A method of describing how a material yields during plastic deformation

Large deformation problem – see large displacement problem

Large displacement problem: Where a structure's displacements are so large that the original stiffness matrix no longer adequately represents the structure. Large displacement problems may result in either small or finite (large) element strains

Line coordinate: A natural coordinate for a one-dimensional element with a value between zero and unity, equal to the ratio of an internal length of the element to its total length

Linear element: Any element with a linear interpolation function

Linear problem: A problem where the material properties, geometry and contact conditions are all linear, ie the stiffness matrix and force vector are not functions of the nodal displacements

Local coordinate integration formulae: A set of integration formulae which can be used to simplify the integration required in the derivation of the element equations

Local coordinate system (in finite element programs): A coordinate system other than the global system, defined to facilitate the generation of a finite element mesh

Local coordinate system (in finite elements): A coordinate system of one particular element, used in the derivation of the element's equations

Lumped mass matrix: One type of mass matrix for a structural element derived by assuming the mass is distributed equally between the nodes (cf consistent mass matrix)

Mass check: A check on a finite element model's mass that is calculated by some programs to provide simple verification that the model is constructed correctly

Mass matrix: A matrix containing details of a body's mass distribution, calculated by assembling the mass matrices of the individual elements. The mass matrix is required in transient dynamics problems

Master degrees of freedom: A limited set of a model's degrees of freedom which characterize the behaviour of the system, and allows a much more economical solution of the system equations

Material non-linearity: Where the material behaviour is not governed by a linear stress–strain relationship (in stress problems), such as those materials considered in plasticity and creep problems

Material property matrix: The [D] matrix, which contains the Young's modulus and Poisson's ratio details in stress problems and the thermal conductivities in thermal problems

Mesh density: The distribution of the elements through a model or part of a model. The mesh density should vary in a model, with the highest density in areas where the field variable is expected to show the greatest variation

Mesh generation: The development and specification of the node and element details, which may be performed either manually or automatically by the computer

Mesh refinement: A reduction in the size of the elements in areas where a rapid change in the field variable is expected

Mid-side node: A node on the side or edge of an element rather than one at a corner

Minimization of a functional: Differentiation of a function derived from the differential equation and boundary conditions of a problem to derive the finite element equations for the problem

Modal analysis: The natural frequency analysis of a body or structure

Modal superposition: A process whereby a number of the mode shapes of a model are factored and combined to predict the dynamic behaviour of a problem

Mode shape: The shape of a structure as it vibrates at a particular frequency

Model equilibrium check: Used as a simple check to show that the loads and constraints are applied correctly. If the problem is in equilibrium, the applied loads should equal the calculated reactions

Multiplex element: A higher order element with sides parallel to the coordinate system in which it is defined. Multiplex elements have quadrilateral and hexahedral shapes

NAFEMS: National Agency for Finite Element Methods and Standards

Natural coordinate: A local coordinate in an element which can vary between zero and unity, or minus unity to unity, depending on the type of element

Natural frequency: A frequency at which an undamped body will vibrate indefinitely if disturbed from its equilibrium position

Newmark-β method: A direct integration method of solution for transient problems

Nodal force: A point load applied at a node

Node: The points of connection between elements, and the location at which the values of the field variable are sought

Non-conforming element: An element that satisfies only the completeness convergence condition, but not the continuity condition

Non-follower loads: Loads which do not maintain the same relative orientation to a body as it undergoes large deformation

Non-linear problem: A problem where the material properties, geometry and/or contact conditions are not linear, ie the stiffness matrix and force vector are functions of the nodal displacements

Non-symmetric loading: Loading which is not symmetric about a given axis or plane

Numerical integration: Numerical evaluation of an integral, rather than analytical integration, which is used where exact, analytical integration is not possible. In particular, Gauss quadrature is most commonly used in the finite element method and produces the exact answer if applied correctly

Orthotropic material: A material with symmetric mechanical properties in two perpendicular directions

p-refinement: An improvement in the accuracy of a finite element mesh by increasing the order of the interpolation function used in the elements

Parametric modelling: Specification of a finite element model using parameters which can be assigned different values as required. This technique is used particularly with automatic mesh generators where the solid model can be defined parametrically

Patch test: The analysis of a small assembly of elements subjected to a constant stress field, to verify that the elements can predict the same constant value

Path plot: A computer generated plot showing the variation of a result along a specified path or line through a model

Pin-jointed element: An element whose nodes (joints) can only transmit forces, and not bending moments

Planar symmetry: A symmetry condition where the same pattern is seen mirrored in a plane

Plane strain: A two-dimensional state of strain where the out-of-plane strain is zero

Plane stress: A two-dimensional state of stress where the out-of-plane stress is zero

Plate element: A flat element which can transmit both bending and membrane loads

Post-processor: A program that presents the results of a finite element analysis, usually graphically, and invariably performs further calculations on the results as required by the user

Potential energy: The energy stored in a body or structure when it is loaded. The potential energy equals the strain energy stored in the material less the work done by the loads

Prandtl's stress function: A stress function used in the analysis of torsion problems

Prandtl-Reuss flow: An incremental theory of plasticity, where the strain is divided into a recoverable elastic component and a non-recoverable plastic component that causes the permanent plastic deformation. The plastic strain is assumed to cause no change in volume

Pre-processor: A program that helps in the generation of the finite element mesh, and prepares the data for direct input into the analysis phase

Prescribed displacement – see enforced displacement

Pressure force: A distributed force applied in stress problems, that acts over the edges or sides of the elements

Program validation: The use of tests, such as single element, patch and benchmark tests, to confirm not only that the elements are formulated correctly, but also that the algorithms are implemented accurately and the pre- and post-processors are reliable and accurate

Quadratic element: Any element with a quadratic interpolation function

Reaction: The force in stress problems, or heat in thermal problems that must be supplied at a constraint to maintain the specified condition at that point

Reaction force check: A check that the sum of the reactions predicted by a finite element analysis equals the sum of the applied loads, as a means of confirming that the model is constructed and loaded correctly

Real constant: A property of an element required by a finite element program. For example, the thickness of a plane stress element, or a second moment of area of a beam

Reduced integration: Approximate numerical integration by the use of fewer integration points than required to calculate the exact answer. Reduced integration leads to cost savings in an analysis

Reducible net/mesh: A series of increasingly fine meshes, where all the coarser meshes are contained in the finer meshes. The meshes are the ideal way to confirm the convergence of a model by h-refinement

Refinement (of the mesh) – see mesh refinement

Repetitive symmetry: A symmetry condition where the same pattern is seen to be repeated down the length of a body

Residual: In the weighted residual method, it is the difference between the exact solution of a differential equation and a solution using an approximating function

Response spectra: Graphs showing the response of a single degree of freedom system (with a range of natural frequencies and damping ratios) to a specified shock load

Response spectrum analysis – see shock spectrum analysis

Restraint – see constraint

Rotational degree of freedom: A variable used to describe the rotation of a discrete point or node in a body. If a node has rotational degrees of freedom it can transmit bending moments

Round-off error: An error occurring in computers because they can only work to a limited number of significant places

Sampling point – see integration point

Section plot: A computer generated plot of a (user-specified) cross-section of a model

Shape function: When a field variable is approximated in an element, its value at any point is expressed in terms of the nodal values. The shape functions dictate the size of these nodal contributions. One shape function is associated with each node of the element

Shape function matrix: The [N] matrix of shape functions, which contains information on the distribution of the unknown field variables in an element in terms of the nodal values

Shell element: Similar to a plate element, except the mid-surface of the element is curved, and as a result the bending and membrane effects are coupled

Shock spectrum analysis: The analysis of a component or structure when subjected to an arbitrary foundation shock load

Shrunken element plot: A plot of the elements produced by the computer where the elements' size is reduced so that the edges of each element are visible, allowing the mesh to be checked

Simplex element: An element with a linear interpolation function, with basic shapes of a line, triangle and tetrahedral in one-, two- and three-dimensions respectively

Single element test: The analysis of a single element subjected to a constant stress field, to verify that the element can predict a constant value

Singularity (in the solution): An error in the solution of the system equations, which occurs when an indeterminate or non-unique solution is possible

Skew distortion: Distortion where the sides of the element are skewed over

Skew symmetry – see anti-symmetry

Slave degrees of freedom: Those degrees of freedom which are not the master degrees of freedom, ie the slaves are not needed to characterize the behaviour of the system

Small strain: When the strain experienced by an element is negligible and the stiffness matrix is a constant. It is possible to have small element strains in a large displacement problem, in which case the stiffness matrix just needs to be reorientated

Solid model: A geometric model of a problem generated by the user from basic entities such as points, lines, areas and volumes, which is subsequently meshed by the finite element program

Solution of the system equations: The phase of the analysis where the equations describing the whole model are solved to yield the values of the field variable at the nodes. Gaussian elimination and the wavefront solution method are commonly used

Space truss: A structure comprised solely of pin-jointed elements in two- or three-dimensions

Static condensation: The process whereby the slave degrees of freedom are removed from the system equations to leave the master degrees of freedom

Steady state problem: A problem which is not a function of time

Stiffness matrix: The [k] matrix, which relates force to displacement in stress problems and applied heat to temperature in thermal problems

Strain energy: The energy stored in a material when it is stressed, and which can be recovered as the loading is removed

Stress averaging: Where the stresses predicted in adjacent elements are averaged at common nodes

Stress pass: The calculation of the displacements of the slave degrees of freedom and element stresses in a dynamics analysis

Stress stiffening: The stiffening (or weakening) of a structure due to the state of stress in the structure

Structural damping matrix: The [C] matrix, which contains information on the viscous damping occurring in the elements in dynamics problems

Submodelling: A technique where a problem is modelled with a coarse mesh, and the results are then applied to a fine mesh of a small detail in the problem

Subparametric element: An element where the order of the geometry interpolation function is less than that of the field interpolation function

Substructuring: The analysis of a large or complex structure by breaking it down into a number of smaller substructures or super-elements

Super-element: A collection of standard elements which is used to represent a part or substructure of a larger complex problem. The super-element is defined by the behaviour of the nodes around its boundary, and is used in a similar way to other finite elements

Superparametric element: An element where the order of the geometry interpolation function is greater than that of the field interpolation function

Surface: Any face of an element. For example, the perimeter and ends of a one-dimensional element, and the face or edges of a two-dimensional element

Swelling: The volumetric enlargement of a material due to neutron bombardment or other effects

Symmetry approximation: The approximation of the true three-dimensional nature of a problem by taking account of symmetry, so that only the repeated pattern is modelled

System equations: The complete assembly of element equations which describes the behaviour of the system being modelled

Taper distortion: A form of element distortion where the sides of the element are tapered to a point

Tension-only element: An element which can only transmit a tensile force, used for example to model cables

Thermal strain: The strain produced in an element due to a change in temperature

Thick beam/plate/shell element: Elements that are formulated to model all bending effects, including the transverse shear deformation

Thin beam/plate/shell element: Elements that are formulated to model bending effects, but assume there is zero transverse shear deformation, ie planes normal to the mid-surface of the element remain normal after loading

Time domain solution: A method of solution of transient dynamic problems where the input is specified as a time history and divided into a number of impulses which are integrated over time

Transient response analysis: The analysis of a structure or component subjected to time varying loads, which could be forces that vary with time, or time functions of displacement, velocity or acceleration

Transient thermal analysis: A thermal analysis where the heat flow and temperatures are a function of time

Transition element: A special element used to connect elements formulated with different order interpolation functions

Translational degree of freedom: A variable used to describe the displacement of a discrete point or node in a body. If a node has translational degrees of freedom, then it can transmit direct forces

Unlimited rigid body motion: Unlimited rigid body motion occurs when a body is not adequately constrained, ie the displacements tend to infinity

Unused node: A node that is specified by the user, but not used in the definition of an element. (It might be used in the definition of a coordinate system)

Validity: How faithfully the physical problem is represented in the computer, which depends on the approximations in the geometry, material properties, loading conditions and constraint conditions

Variational formulation: A method of deriving the finite element equations for a problem by minimizing a functional

Vector plot: A computer generated plot of the distribution of a vector quantity. The values are presented as small arrows drawn over the model, where the lengths and/or colours of the arrows indicate the magnitude, and their orientation shows the direction. Used for example with principal stresses and heat flows

Volume check: A check on a finite element model's volume that is calculated by some programs, to provide simple verification that the model is constructed correctly

Volume coordinate: A natural coordinate for a tetrahedral element with a value between zero and unity, equal to the ratio of a tetrahedral portion of the element to its total volume

Von-Mises failure criterion: A method of predicting when a material can be considered to have failed

Warpage angle: A measure of warping distortion

Warping distortion: A form of element distortion where the nodes of a quadrilateral (or face of a solid element) do not lie in the same plane

Wavefront solution method: A method of solution for the nodal values of the unknown field variable where the complete system equations are never assembled. As the assembly of the element equations proceeds, the degrees of freedom that are not used by the remaining elements are 'eliminated' from the equations

Weighted residual method: A method of calculating the finite element equations of a problem directly from its governing differential equation

Weighting function (in Gauss quadrature): A constant used to factor a value calculated at a sampling point when the element equations are evaluated during numerical integration

Weighting function (in the weighted residual method): A function used to weight the residual in the derivation of the finite element equations. In Galerkin's method the weighting functions equal the shape functions

Index

A simple, graphics based, user friendly finite element program is available for PC computers to accompany this book. The program has been developed to explore and present the inner workings of the finite element method rather than perform complex analyses, but nevertheless can be used for the analysis of realistic problems. The essential feature of the program is a 'transparent' mode, whereby the student can selectively follow the calculations of a complete analysis, and examine the details of individual elements and their performances.

For further details photocopy and complete the slip below and send it to the author.

To   Dr M J Fagan,
     Department of Engineering Design and Manufacture,
     University of Hull,
     Cottingham Road,
     HULL,
     N Humberside.
     HU6 7RX

Please send further information of the program to supplement the book 'Finite Element Analysis – Theory and Practice'.

Name: _____

Address: _____
           _____
           _____
           _____
           _____